Synchronized Phasor Measurements
and Their Applications

Power Electronics and Power Systems

Series Editors: M. A. Pai   Alex Stankovic
                 University of Illinois at Urbana-Champaign   Northeastern University
                 Urbana, Illinois   Boston, Massachusetts

Continued after index

A.G. Phadke · J.S. Thorp

Synchronized Phasor Measurements and Their Applications

 Springer

A.G. Phadke
Virginia Polytechnic Institute and
State University
Blacksburg, VA
USA

J.S. Thorp
Virginia Polytechnic Institute and
State University
Blacksburg, VA
USA

Series Editors
M.A. Pai
University of Illinois at Urbana-Champaign
Urbana, IL
USA

Alex Stankovic
Northeastern University
Boston, MA
USA

ISBN: 978-0-387-76535-8   e-ISBN: 978-0-387-76537-2
DOI: 10.1007/978-0-387-76537-2

Library of Congress Control Number: 2008921736

Printed on acid-free paper

9 8 7 6 5 4 3 2 1

springer.com

Preface

Synchronized phasor measurements have become the measurement technique of choice for electric power systems. They provide positive sequence voltage and current measurements synchronized to within a microsecond. This has been made possible by the availability of Global Positioning System (GPS) and the sampled data processing techniques developed for computer relaying applications. In addition to positive-sequences voltages and currents, these systems also measure local frequency and rate of change of frequency, and may be customized to measure harmonics, negative and zero sequence quantities, as well as individual phase voltages and currents. At present there are about 24 commercial manufacturers of phasor measurement units (PMUs), and industry standards developed in the Power System Relaying Committee of IEEE has made possible the interoperability of units from different manufacturers.

Recent spate of spectacular blackouts on power systems throughout the world has provided an added impetus to widescale deployment of PMUs. Positive sequence measurements provide the most direct access to the state of the power system at any given instant. Many applications of these measurements have been discussed in the technical literature, and no doubt many more applications will be developed in coming years.

The authors have been associated with this technology since its birth, and they and their colleagues and students have produced a rich body of literature on the subject of phasor measurement technology and its applications. Other researchers around the world have also made significant contributions to the field. Our aim in writing this book is to present to the interested reader a coherent account of the development of the technology, and of the emerging applications of these measurements. It is our hope that this book will help power system engineers understand the basics of "synchronized phasor measurement systems". This technology is bound to usher in an era of improved monitoring, protection, and control of power systems.

Blacksburg, Virginia
January, 2008

Arun G. Phadke
James S. Thorp

Contents

Part I: Phasor Measurement Techniques

Chapter 1 Introduction

1.1 Historical overview

Phase angles of voltage phasors of power network buses have always been of special interest to power system engineers. It is well-known that active (real) power flow in a power line is very nearly proportional to the sine of the angle difference between voltages at the two terminals of the line. As many of the planning and operational considerations in a power network are directly concerned with the flow of real power, measuring angle differences across transmission has been of concern for many years. The earliest modern application involving direct measurement of phase angle differences was reported in three papers in early 1980s[1,2,3]. These systems used LORAN-C, GOES satellite transmissions, and the HBG radio transmissions (in Europe) in order to obtain synchronization of reference time at different locations in a power system. The next available positive-going zero-crossing of a phase voltage was used to estimate the local phase angle with respect to the time reference. Using the difference of measured angles on a common reference at two locations, the phase angle difference between voltages at two buses was established. Measurement accuracies achieved in these systems were of the order of 40 μs. Single-phase voltage angles were measured and, of course no attempt was made to measure the prevailing voltage phasor magnitude. Neither was any account taken of the harmonics contained in the voltage waveform. These methods of measuring phase angle differences are not suitable for generalization for wide-area phasor measurement systems, and remain one-of-a-kind systems which are no longer in use.

The modern era of phasor measurement technology has its genesis in research conducted on computer relaying of transmission lines. Early work on transmission line relaying with microprocessor-based relays showed that the available computer power in those days (1970s) was barely sufficient to manage the calculations needed to perform all the transmission line relaying functions.

A significant portion of the computations was dedicated to solving six fault loop equations at each sample time in order to determine if any one of the ten

A.G. Phadke, J.S. Thorp, *Synchronized Phasor Measurements and Their Applications*,
DOI: 10.1007/978-0-387-76537-2_1, © Springer Science+Business Media, LLC 2008

types of faults possible on a three-phase transmission line are present. The search for methods which would eliminate the need to solve the six equations finally yielded a new relaying technique which was based on symmetrical component analysis of line voltages and currents. Using symmetrical components, and certain quantities derived from them, it was possible to perform all fault calculations with a single equation. In a paper published in 1977 [4] this new symmetrical component-based algorithm for protecting a transmission line was described. As a part of this theory, efficient algorithms for computing symmetrical components of three-phase voltages and currents were described, and the calculation of positive-sequence voltages and currents using the algorithms of that paper gave an impetus for the development of modern phasor measurement systems. It was soon recognized that the positive-sequence measurement (a part of the symmetrical component calculation) is of great value in its own right. Positive-sequence voltages of a network constitute the state vector of a power system, and it is of fundamental importance in all of power system analysis. The first paper to identify the importance of positive-sequence voltage and current phasor measurements, and some of the uses of these measurements, was published in 1983 [5], and this last paper can be viewed as the starting point of modern synchronized phasor measurement technology. The Global Positioning System (GPS) [6] was beginning to be fully deployed around that time. It became clear that this system offered the most effective way of synchronizing power system measurements over great distances. The first prototypes of the modern "phasor measurement units" (PMUs) using GPS were built at Virginia Tech in early 1980s, and two of these prototypes are shown in Figure 1.1. The prototype PMU units built at Virginia Tech were deployed at a few substations of the Bonneville Power Administration, the American Electric Power Service Corporation, and the New York Power Authority. The first commercial manufacture of PMUs with Virginia Tech collaboration was started by Macrodyne in 1991 [7]. At present, a number of manufacturers offer PMUs as a commercial product, and deployment of PMUs on power systems is being carried out in earnest in many countries around the world. IEEE published a standard in 1991 [8] governing the format of data files created and transmitted by the PMU. A revised version of the standard was issued in 2005.

Concurrently with the development of PMUs as measurement tools, research was ongoing on applications of the measurements provided by the PMUs. These applications will be discussed in greater detail in later chapters of this book. It can be said now that finally the technology of synchronized phasor measurements has come of age, and most modern power systems around the world are in the process of installing wide-area measurement systems consisting of the phasor measurement units.

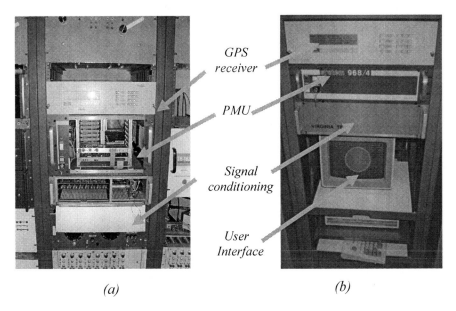

(a)   *(b)*

Fig. 1.1 The first phasor measurement units (PMUs) built at the Power Systems Research Laboratory at Virginia Tech. The GPS receiver clock was external to the PMU, and with the small number of GPS satellites deployed at that time, the clock had to be equipped with a precision internal oscillator which maintained accurate time in the absence of visible satellites.

1.2 Phasor representation of sinusoids

Consider a pure sinusoidal quantity given by

$$x(t) = X_\mathrm{m} \cos(\omega t + \phi) \tag{1.1}$$

ω being the frequency of the signal in radians per second, and ϕ being the phase angle in radians. X_m is the peak amplitude of the signal. The root mean square (RMS) value of the input signal is $(X_\mathrm{m}/\sqrt{2})$. Recall that RMS quantities are particularly useful in calculating active and reactive power in an AC circuit.

Equation (1.1) can also be written as

$$x(t) = Re\{X_\mathrm{m}\, e^{j(\omega t + \phi)}\} = Re[\{e^{j(\omega t)}\}\, X_\mathrm{m}\, e^{j\phi}].$$

It is customary to suppress the term $e^{j(\omega t)}$ in the expression above, with the understanding that the frequency is ω. The sinusoid of Eq. (1.1) is represented by a complex number X known as its phasor representation:

$$x(t) \leftrightarrow X = (X_\mathrm{m}/\sqrt{2})\, e^{j\phi} = (X_\mathrm{m}/\sqrt{2})\, [\cos \phi + j \sin \phi]. \tag{1.2}$$

A sinusoid and its phasor representation are illustrated in Figure 1.2.

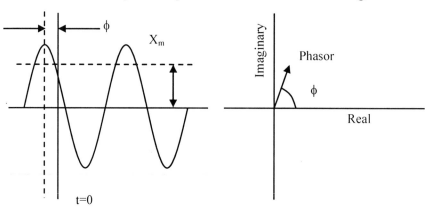

(a)   (b)

Fig. 1.2 A sinusoid (**a**) and its representation as a phasor (**b**). The phase angle of the phasor is arbitrary, as it depends upon the choice of the axis $t = 0$. Note that the length of the phasor is equal to the RMS value of the sinusoid.

It was stated earlier that the phasor representation is only possible for a pure sinusoid. In practice a waveform is often corrupted with other signals of different frequencies. It then becomes necessary to extract a single frequency component of the signal (usually the principal frequency of interest in an analysis) and then represent it by a phasor. Extracting a single frequency component is often done with a "Fourier transform" calculation. In sampled data systems, this becomes the "discrete Fourier transform" (DFT) or the "fast fourier transform" (FFT). These transforms are reviewed in the next section. The phasor definition also implies that the signal is unchanging for all time. However, in all practical cases, it is only possible to consider a portion of time span over which the phasor representation is considered. This time span, also known as the "data window", is very important in phasor estimation of practical waveforms. It will be considered in greater detail in later sections.

1.3 Fourier series and Fourier transform

1.3.1 Fourier series

Let $x(t)$ be a periodic function of t, with a period equal to T. Then $x(t+kT) = x(t)$ for all integer values of k. A periodic function can be expressed as a Fourier series:

$$x(t) = \frac{a_0}{2} + \sum_{k=1}^{\infty} a_k \cos(\frac{2\pi kt}{T}) + \sum_{k=1}^{\infty} b_k \sin(\frac{2\pi kt}{T}), \qquad (1.3)$$

where the constants a_k and b_k are given by

$$a_k = \frac{2}{T} \int_{-T/2}^{+T/2} x(t) \cos(\frac{2\pi kt}{T}) dt, \quad k = 0, 1, 2, \cdots,$$

$$b_k = \frac{2}{T} \int_{-T/2}^{+T/2} x(t) \sin(\frac{2\pi kt}{T}) dt, \quad k = 1, 2, \cdots. \qquad (1.4)$$

The Fourier series can also be written in the exponential form

$$x(t) = \sum_{k=-\infty}^{\infty} \alpha_k e^{\frac{j2\pi kt}{T}} \qquad (1.5)$$

with

$$\alpha_k = \frac{1}{T} \int_{-T/2}^{+T/2} x(t) e^{-\frac{j2\pi kt}{T}} dt, \quad k = 0, \pm1, \pm2, \cdots. \qquad (1.6)$$

Note that the summation in Eq. (1.5) goes from $-\infty$ to $+\infty$, while the summations in Eq. (1.3) go from 1 to $+\infty$. The change in summation limits is accomplished by noting that the cosine and sine functions are even and odd functions of k, and thus expanding the summation limits to ($-\infty$ to $+\infty$) and removing the factor 2 in front of the integrals for a_k and b_k leads to the desired exponential form of the Fourier series.

Example 1.1
Consider a periodic square wave signal with a period T as shown in Figure 1.3. This is an even function of time. The Fourier coefficients (in exponential form) are given by

$$\alpha_k = \frac{1}{T} \int_{-T/4}^{+T/4} e^{-\frac{j2\pi kt}{T}} dt, \quad k = 0, \pm1, \pm2, \cdots$$

$$= \frac{1}{\pi k} \sin(\frac{k\pi}{2}).$$

Hence $\alpha_0 = 1/2$,
$\alpha_1 = 1/\pi, \alpha_{-1} = 1/\pi,$
$\alpha_3 = -1/3\pi, \alpha_{-3} = -1/3\pi,$
$\alpha_5 = 1/5\pi, \alpha_{-5} = 1/5\pi,$ etc., and all even coefficients are zero.

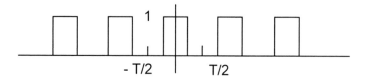

Fig. 1.3 A square wave function with a period T, with duty cycle equal to half, with the $t = 0$ axis so chosen that the function is an even function.

Thus, the Fourier series of the square wave signal is

$$x(t) = \frac{1}{2} + \frac{2}{\pi} \left[\cos(\frac{2\pi t}{T}) - \frac{1}{3}\cos(\frac{6\pi t}{T}) + \frac{1}{5}\cos(\frac{10\pi t}{T}) - \cdots \right].$$

The sum of first seven terms of the series is shown in Figure 1.4 below:

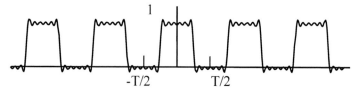

Fig. 1.4 A square wave approximated by 7 terms of the Fourier series. With more terms, the waveform approaches the square shape. The oscillations are known as the Gibbs phenomenon and are inescapable when step functions are approximated by the Fourier series.

1.3.2 Fourier transform

There are several excellent text books devoted to the subject of Fourier transforms [9,10]. The reader should consult those books for a more complete account of the Fourier transform theory. Here we present only those topics which are of direct interest for phasor estimation in power system applications.

The Fourier transform of a continuous time function $x(t)$ satisfying certain integrability conditions [9] is given by

$$X(f) = \int_{-\infty}^{+\infty} x(t)e^{-j2\pi ft}\,dt \qquad (1.7)$$

and the inverse Fourier transform recovers the time function from its Fourier transform:

$$x(t) = \int_{-\infty}^{+\infty} X(f)e^{j2\pi ft}df. \tag{1.8}$$

An important function frequently used in calculations using sampled data is the impulse function $\delta(t)$ defined by

$$x(t_0) = \int_{-\infty}^{+\infty} \delta(t - t_0)x(t)dt. \tag{1.9}$$

The impulse function (also known as a distribution or a Dirac delta function) is a sampling function in the sense that when the integration in Eq. (1.9) is performed, the result is the sampled value of the function $x(t)$ at $t = t_0$. The integrals of the type shown in Eq. (1.9) are known as convolutions. Thus the sampling process at uniform intervals ΔT apart can be considered to be a convolution of the input signal and a string of impulse functions $\delta(t - k\Delta T)$, where k ranges from $-\infty$ to $+\infty$.

The convolutions of two time functions and their Fourier transforms have a convenient relationship. Consider the convolution $z(t)$ of two time functions $x(t)$ and $y(t)$:

$$z(t) = \int_{-\infty}^{+\infty} x(\tau)y(\tau - t)d\tau \equiv x(t) * y(t). \tag{1.10}$$

The important result regarding convolutions is the following property:

Property 1 *The Fourier transform of a convolution is equal to the product of the Fourier transform of the functions being convolved, or:*
If $s(t) = x(t)*y(t)$, then $S(f) = X(f).Y(f)$
and similarly, the inverse Fourier transform of a convolution of two Fourier transforms is a product of the corresponding inverse Fourier transforms:
If $Z(f) = X(f)*Y(f)$, then $z(t) = x(t).y(t)$.

Next we illustrate the second of the above two statements. Consider the functions $x(t) = \cos(\omega t)$ and $y(t) = \sin(\omega t)$, with $\omega = 2\pi f_0$. The Fourier transforms of $x(t)$ and $y(t)$ are

$$X(f) = \int_{-\infty}^{+\infty} \cos(2\pi f_0 t)e^{-j2\pi ft}dt = \int_{-\infty}^{+\infty} \frac{e^{-j2\pi(f-f_0)t} + e^{-j2\pi(f+f_0)t}}{2}dt$$

$$= \frac{1}{2}[\delta(f - f_0) + \delta(f + f_0)]$$

and similarly

$$Y(f) = \frac{j}{2}[\delta(f+f_0) - \delta(f-f_0)]$$

The Fourier transforms of a pure cosine wave of unit amplitude is a pair of real impulse functions in frequency domain located at $\pm f_0$ and that of a pure sine wave of unit amplitude is a pair of imaginary impulse functions of opposite signs at $\pm f_0$.

The convolution of the two Fourier transforms determined above in the frequency domain is

$$S(f) = \int_{-\infty}^{+\infty} \frac{1}{2}[\delta(\phi-f_0) + \delta(\phi+f_0)]\frac{j}{2}[\delta(f+f_0-\phi) - \delta(f-f_0-\phi)]d\phi$$

$$= \frac{j}{4}\int_{-\infty}^{+\infty}[\delta(\phi-f_0)\delta(f+f_0-\phi) + \delta(\phi+f_0)\delta(f+f_0-\phi)$$

$$- \delta(\phi-f_0)\delta(f-f_0-\phi) - \delta(\phi+f_0)\delta(f-f_0-\phi)]d\phi .$$

Using the sampling property of the integrals involving impulse functions

$$S(f) = \frac{j}{4}[\delta(f+2f_0) - \delta(f-2f_0)]$$

the inverse Fourier transform of $S(f)$ is clearly

$$s(t) = \frac{1}{2}\sin(4\pi ft) = \sin(2\pi ft)\cos(2\pi ft) = x(t).y(t)$$

This property of convolutions will be used in discussing the sampling process and the DFT. Some other properties of the Fourier transform which are particularly useful in our development are stated next with accompanying examples.

Property 2 *The Fourier transform of an even function is an even function of frequency. If the even function is real, the Fourier transform is also real and even.*

Consider an even function of time, $x(t)$, so that $x(-t) = x(t)$. Let $x(t)$ be complex, $x(t) = r(t) + js(t)$. The Fourier transform $X(f)$ of this function is given by

$$X(f)= \int_{-\infty}^{+\infty}x(t)e^{-j2\pi ft}\,dt = \int_{-\infty}^{+\infty}r(t)e^{-j2\pi ft}\,dt + j \int_{-\infty}^{+\infty}s(t)e^{-j2\pi ft}\,dt =$$

$$= \int_{-\infty}^{+\infty}r(t)\cos(2\pi ft)\,dt + j \int_{-\infty}^{+\infty}r(t)\sin(2\pi ft)\,dt + j \int_{-\infty}^{+\infty}s(t)\cos(2\pi ft)\,dt - \int_{-\infty}^{+\infty}s(t)\sin(2\pi ft)\,dt$$

The second and fourth integrals are zero, since the integrands are odd functions of time. Thus

$$X(f)= \int_{-\infty}^{+\infty}r(t)\cos(2\pi ft)\,dt + j \int_{-\infty}^{+\infty}s(t)\cos(2\pi ft)\,dt$$

Since $\cos(2\pi ft) = \cos(-2\pi ft)$, it follows that $X(f) = X(-f)$.
Also, if $x(t)$ is real $= r(t)$, the Fourier transform of $x(t)$ is $R(f)$, which is real and even.

Property 3 *The Fourier transform of an odd function is an odd function of frequency. If the odd function is real, the Fourier transform is imaginary and odd.*

Consider an odd function of time, $x(t)$, so that $x(-t) = - x(t)$. Let $x(t)$ be complex, $x(t) = r(t) + js(t)$. The Fourier transform $X(f)$ of this function is given by

$$X(f)= \int_{-\infty}^{+\infty}x(t)e^{-j2\pi ft}\,dt = \int_{-\infty}^{+\infty}r(t)e^{-j2\pi ft}\,dt + j \int_{-\infty}^{+\infty}s(t)e^{-j2\pi ft}\,dt =$$

$$= \int_{-\infty}^{+\infty}r(t)\cos(2\pi ft)\,dt + j \int_{-\infty}^{+\infty}r(t)\sin(2\pi ft)\,dt + j \int_{-\infty}^{+\infty}s(t)\cos(2\pi ft)\,dt - \int_{-\infty}^{+\infty}s(t)\sin(2\pi ft)\,dt .$$

The first and third integrals are zero, since the integrands are odd functions of time. Thus

$$X(f)= j \int_{-\infty}^{+\infty}r(t)\sin(2\pi ft)\,dt - \int_{-\infty}^{+\infty}s(t)\sin(2\pi ft)\,dt .$$

Since $\sin(2\pi ft) = -\sin(-2\pi ft)$, it follows that $X(f) = - X(-f)$.
Also, if $x(t)$ is real $= r(t)$, the Fourier transform of $x(t)$ is $jR(f)$, which is imaginary and odd.

Property 4 *The Fourier transform of a real function has an even real part and an odd imaginary part.*

Consider a real function of time $x(t) = r(t) + j0$. The Fourier transform is given by

$$X(f) = \int_{-\infty}^{+\infty} r(t)\cos(2\pi ft)dt + j\int_{-\infty}^{+\infty} r(t)\sin(2\pi ft)dt = R_1(f) + jR_2(f)$$

Since the cosine and sine functions are respectively even and odd functions of frequency, it is clear that $R_1(f)$ is an even function, and $R_2(f)$ is an odd function of frequency.

Property 5 *The Fourier transform of a periodic function is a series of impulse functions of frequency.*

If $x(t)$ is a periodic function of t, with a period equal to T, it can be expressed as an exponential Fourier series given by Eqs. (1.5) and (1.6):

$$x(t) = \sum_{k=-\infty}^{\infty} \alpha_k e^{\frac{j2\pi kt}{T}}$$

with

$$\alpha_k = \frac{1}{T}\int_{-T/2}^{+T/2} x(t)e^{-\frac{j2\pi kt}{T}} dt, \quad k = 0, \pm 1, \pm 2, \ldots$$

The Fourier transform of $x(t)$ expressed in the exponential form is given by

$$X(f) = \int_{-\infty}^{+\infty} x(t)e^{-j2\pi ft}dt = \int_{-\infty}^{+\infty}[\sum_{k=-\infty}^{\infty}\alpha_k e^{\frac{j2\pi kt}{T}}]e^{-j2\pi ft}dt = \sum_{k=-\infty}^{\infty}\int_{-\infty}^{+\infty}\alpha_k e^{\frac{j2\pi kt}{T}}e^{-j2\pi ft}dt$$

$$= \sum_{k=-\infty}^{\infty}\int_{-\infty}^{+\infty}\alpha_k e^{j2\pi t\left\{\frac{k}{T}-f\right\}} dt$$

where the order of the summation and integration has been reversed (assuming that this is permissible). Setting $f_0 = 1/T$, the fundamental frequency of the periodic signal, the integral of the exponential term in the last form is the impulse function $\delta(kf_0 - f)$, and thus the Fourier transform of periodic $x(t)$ is

$$X(f) = \sum_{k=-\infty}^{\infty} \alpha_k\, \delta(f - \frac{k}{T}), \text{ with}$$

$$\alpha_k = \frac{1}{T} \int_{-T/2}^{+T/2} x(t)e^{-\frac{j2\pi kt}{T}}\, dt, \quad k = 0,\pm1,\pm2,\ldots$$

These are a series of impulses located at multiples of the fundamental frequency f_0 of the periodic signal with impulse magnitudes being equal to amplitude of each frequency component in the input signal.

Property 6 *The Fourier transform of a series of impulses is a series of impulse functions in the frequency domain.*

Consider the function

$$x(t) = \sum_{k=-\infty}^{\infty} \delta(t - kT)$$

This is a periodic function with period T. Hence its Fourier transform (by Property 5 above) is

$$X(f) = \sum_{k=-\infty}^{\infty} \alpha_k\, \delta(f - \frac{k}{T}), \text{ with}$$

$$\alpha_k = \frac{1}{T} \int_{-T/2}^{+T/2} \delta(t)e^{-\frac{j2\pi kt}{T}}\, dt, \quad k = 0,\pm1,\pm2,\cdots$$

Since the delta function in the integrand produces a sample of the exponent at $t = 0$, α_k is equal to $1/T$ for all k, and the Fourier transform of $x(t)$ becomes

$$X(f) = \frac{1}{T} \sum_{k=-\infty}^{\infty} \delta(f - \frac{k}{T}),$$

which is a pulse train in the frequency domain at intervals kf_0 and a magnitude of $1/T$.

Example 1.2
Consider a rectangular input signal as shown in Figure 1.5. This is an even function of time.

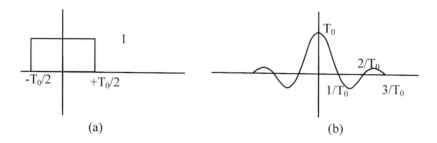

Fig. 1.5 (a) A rectangular function of time with the $t = 0$ axis so chosen that the function is an even function. The duration of the signal is $2T_0$. **(b)** Fourier transform of the function.

The Fourier transform of this time function is given by

$$X(f) = \int_{-\infty}^{+\infty} x(t)e^{-j2\pi ft}\,dt = \int_{t_1}^{t_1+T_0} e^{-j2\pi ft}\,dt = e^{j2\pi f(t_1+\frac{T_0}{2})}T_0\frac{\sin(2\pi f\frac{T_0}{2})}{(2\pi f\frac{T_0}{2})}.$$

The first term in the Fourier transform is a phase shift factor and has been omitted from the plot in Figure 1.5b for convenience. If the rectangular wave is centered at the origin, $t_1 = -T_0/2$, and the phase shift factor vanishes. This is also in keeping with Property 2 of the Fourier transform given above, which states that the Fourier transform of a real even function must be real and even function of frequency.

1.4 Sampled data and aliasing

Sampled data from input signals are the starting point of digital signal processing. The computation of phasors of voltages and currents begins with samples of the waveform taken at uniform intervals $k\Delta T$, ($k = 0, \pm1, \pm2, \pm3, \pm4, \cdots$). Consider an input signal $x(t)$ which is being sampled, yielding sampled data $x(k\Delta T)$. We may view the sampled data as a time function $x'(t)$ consisting of uniformly spaced impulses, each with a magnitude $x(k\Delta T)$:

$$x'(t) = \sum_{k=-\infty}^{\infty} x(k\Delta t)\,\delta(t - k\Delta T) \tag{1.11}$$

It is interesting to determine the Fourier transform of the sampled data function given by Eq. (1.11). Note that the sampled data function is a product of the function $x(t)$ and the sampling function $\delta(t - k\Delta T)$, the product being interpreted in the sense of Eq. (1.9). Hence the Fourier transform $X'(f)$ of $x'(t)$ is the convolution of the Fourier transforms of $x(t)$ and of the unit impulse train. By Property 6 of Section 1.3, the Fourier transform of the impulse train is

$$\Delta(f) = \frac{1}{\Delta T} \sum_{k=-\infty}^{\infty} \delta(f - \frac{k}{\Delta T}) \qquad . \qquad (1.12)$$

Hence the Fourier transform of the sampled data function is the convolution of $\Delta(f)$ and $X(f)$

$$X'(f) = \frac{1}{\Delta T} \int_{-\infty}^{+\infty} X(\phi) \sum_{k=-\infty}^{\infty} \delta(f - \frac{k}{\Delta T} - \phi) \, d\phi$$

$$= \frac{1}{\Delta T} \sum_{k=-\infty}^{\infty} \int_{-\infty}^{+\infty} X(\phi)\delta(f - \frac{k}{\Delta T} - \phi) \, d\phi \qquad (1.13)$$

$$= \frac{1}{\Delta T} \sum_{k=-\infty}^{\infty} X(f - \frac{k}{\Delta T})$$

Once again the order of summation and integration has been reversed (it being assumed that this is permissible), and the integral is evaluated by the use of the sampling property of the impulse function.

The relationship between the Fourier transforms of $x(t)$ and $x'(t)$ are as shown in Figure 1.6. The Fourier transform of $x(t)$ is shown to be band-limited, meaning that it has no components beyond a cut-off frequency f_c. The sampled data has a Fourier transform which consists of an infinite train of the Fourier transforms of $x(t)$ centered at frequency intervals of $(k/\Delta T)$ for all k. Recall that the sampling interval is ΔT, so that the sampling frequency $f_s = (1/\Delta T)$.

If the cut-off frequency f_c is greater than one-half of the sampling frequency f_s, the Fourier transform of the sampled data will be as shown in Figure 1.7. In this case, the spectrum of the sampled data is different from that of the input signal in the region where the neighboring spectra overlap as shown by the shaded region in Figure 1.7. This implies that frequency components estimated from the sampled data in this region will be in error, due to a phenomenon known as "aliasing".

It is clear from the above discussion that in order to avoid errors due to aliasing, the bandwidth of the input signal must be less than half the sampling frequency utilized in obtaining the sampled data. This requirement is known as the "Nyquist criterion".

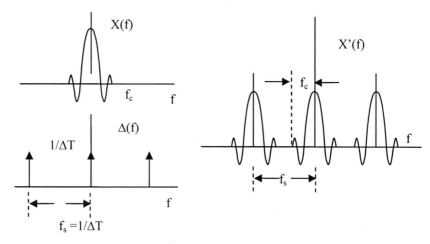

Fig. 1.6 Fourier transform of the sampled data function as a convolution of the Transforms $X(f)$ and $\Delta(f)$. The sampling frequency is f_s, and $X(f)$ is band-limited between $\pm f_c$.

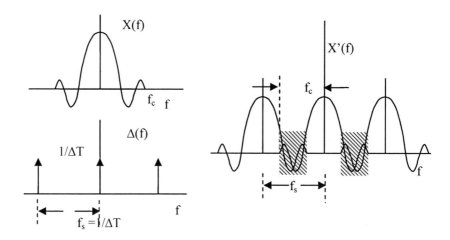

Fig. 1.7 Fourier transform of the sampled data function when the input signal is band-limited to a frequency greater than half the sampling frequency. The estimate of frequencies from sampled data in the shaded region will be in error because of aliasing.

In order to avoid aliasing errors, it is customary in all sampled data systems used in phasor estimation to use anti-aliasing filters which band-limit the input signals to below half the sampling frequency chosen. Note that the signal input cut-off frequency must be *less than* one half the sampling frequency. In practice, the signal is usually band-limited to a value much smaller than the one required for meeting the Nyquist criterion. Anti-aliasing filters are generally passive low-pass R-C filters [11], although active filters may also be used for obtaining a sharp cut-off characteristic. In addition to passive anti-aliasing filters, digital filters may also be used in special cases (e.g., with oversampling and decimation). All anti-aliasing filters introduce frequency-dependent phase shift in the input signal which must be compensated for in determining the phasor representation of the input signal. This will be discussed further in Chapter 5 where the 'Synchrophasor' standard is described.

1.5 Discrete Fourier transform (DFT)

DFT is a method of calculating the Fourier transform of a small number of samples taken from an input signal $x(t)$. The Fourier transform is calculated at discrete steps in the frequency domain, just as the input signal is sampled at discrete instants in the time domain. Consider the process of selecting N samples: $x(k\Delta T)$ with $\{k = 0, 1, 2, \cdots ,N-1\}$, ΔT being the sampling interval. This is equivalent to multiplying the sampled data train by a "windowing function" $w(t)$, which is a rectangular function of time with unit magnitude and a span of $N\Delta T$. With the choice of samples ranging from 0 to $N-1$, it is clear that the windowing function can be viewed as starting at $-\Delta T/2$ and ending at $(N-1/2)\Delta T$. The function $x(t)$, the sampling function $\Delta(t)$, and the windowing function $w(t)$ along with their Fourier transforms are shown in Figure 1.8.

Consider the collection of signal samples which fall in the data window: $x(k\Delta T)$ with $\{k = 0,1,2, \cdots ,N-1\}$. These samples can be viewed as being obtained by the multiplication of the signal $x(t)$, the sampling function $\delta(t)$, and the windowing function $\omega(t)$:

$$y(t) = x(t)\delta(t)w(t) = \sum_{k=0}^{N-1} x(k\Delta T)\delta(t-k\Delta T),\qquad (1.14)$$

where once again the multiplication with the delta function is to be understood in the sense of the integral of Eq. (1.9). The Fourier transform of the sampled windowed function $y(t)$ is then the convolution of Fourier transforms of the three functions.

The Fourier transform of $y(t)$ is to be sampled in the frequency domain in order to obtain the DFT of $y(t)$. The discrete steps in the frequency domain are multiples of $1/T_0$, where T_0 is the span of the windowing function. The frequency sampling function $\Phi(f)$ is given by

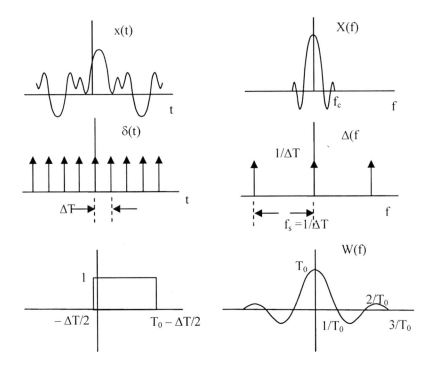

Fig. 1.8 Time functions and Fourier transforms $x(t)$, $\delta(t)$, and $\omega(t)$. Note that once again the phase shift factor from $\Omega(f)$ has been omitted.

$$\Phi(f) = \sum_{n=-\infty}^{\infty} \delta\left(f - \frac{n}{T_0}\right) \qquad (1.15)$$

and its inverse Fourier transform (by Property 6 of Fourier transforms) is

$$\phi(t) = T_0 \sum_{n=-\infty}^{\infty} \delta(t - nT_0). \qquad (1.16)$$

In order to obtain the samples in the frequency domain, we must multiply the Fourier transform $Y(f)$ with $F(f)$. To obtain the corresponding time domain function $x'(t)$ we will require a convolution in the time domain of $y(t)$ and $\phi(t)$:

$$x'(t) = y(t) * \phi(t)$$

$$x'(t) = y(t) * \phi(t) = \left[\sum_{k=0}^{N-1} x(k\Delta T)\delta(t - k\Delta T) \right] * \left[T_0 \sum_{n=-\infty}^{\infty} \delta(t - nT_0) \right]$$

$$= T_0 \sum_{n=-\infty}^{\infty} \left[\sum_{k=0}^{N-1} x(k\Delta T)\delta(t - k\Delta T - nT_0) \right]. \qquad (1.17)$$

This function is periodic with a period T_0. The functions $x(t)$, $y(t)$, and $x'(t)$ are shown in Figure 1.9. The windowing function limits the data to samples 0 through $N-1$, and the sampling in frequency domain transforms the original N samples in time domain to an infinite train of N samples with a period T_0 as shown in Figure 1.9 (c). Note that although the original function $x(t)$ was not periodic, the function $x'(t)$ is, and we may consider this function to be an approximation of $x(t)$.

The Fourier transform of the periodic function $x'(t)$ is a sequence of impulse functions in frequency domain by Property 5 of the Fourier transform. Thus

$$X'(f) = \sum_{n=-\infty}^{\infty} \alpha_n \, \delta(f - \frac{n}{T_0}), \text{ with}$$

$$\alpha_n = \frac{1}{T_0} \int_{-T_0/2}^{T_0-T_0/2} x'(t) e^{-\frac{j2\pi nt}{T_0}} \, dt, \quad n = 0, \pm 1, \pm 2, \cdots.$$

(1.18)

Substituting for $x'(t)$ in the above expression for α_n,

$$\alpha_n = \frac{1}{T_0} \int_{-T_0/2}^{T_0-T_0/2} \left\{ T_0 \sum_{m=-\infty}^{\infty} \left[\sum_{k=0}^{N-1} x(k\Delta T) \delta(t - k\Delta T - mT_0) \right] \right\} e^{-\frac{j2\pi nt}{T_0}} \, dt,$$

$$n = 0, \pm 1, \pm 2, \cdots$$

(1.19)

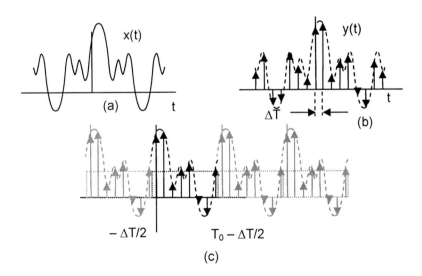

Fig. 1.9 (a) The input function $x(t)$, its samples (b), and (c) the Fourier transform of the windowed function $x'(t)$.

The index m designates the train of periods shown in Figure 1.9 (c). Since the limits on the integration span one period only, we may remove the summation on m, and set $m = 0$, thus using only the samples over the period shown in bold in Figure 1.9 (c). Equation (1.15) then becomes

$$\alpha_n = \int_{-T_0/2}^{T_0-T_0/2} \left[\sum_{k=0}^{N-1} x(k\Delta T)\,\delta(t - k\Delta T) \right] e^{-\frac{j2\pi nt}{T_0}}\, dt, \quad or$$

$$\alpha_n = \sum_{k=0}^{N-1} \int_{-T_0/2}^{T_0-T_0/2} x(k\Delta T)\delta(t - k\Delta T)\, e^{-\frac{j2\pi nt}{T_0}}\, dt, \quad = \sum_{k=0}^{N-1} x(k\Delta T) e^{-\frac{j2\pi nkn\Delta T}{T_0}} \tag{1.20}$$

$$n = 0, \pm 1, \pm 2, \cdots.$$

Since there are N samples in the data window T_0, $N\Delta T = T_0$. Therefore

$$\alpha_n = \sum_{k=0}^{N-1} x(k\Delta T) e^{-\frac{j2\pi kn}{N}}, \; with \; n = 0, \pm 1, \pm 2, \cdots. \tag{1.21}$$

Although the index n goes over all positive and negative integers, it should be noted that there are only N distinct coefficients α_n. Thus, α_{N+1} is the same as α_1 and the Fourier transform $X'(f)$ has only N distinct values corresponding to frequencies $f = n/T_0$, with n ranging from 0 through $N - 1$:

$$X'(\frac{n}{T_0}) = \sum_{k=0}^{N-1} x(k\Delta T) e^{-\frac{j2\pi kn}{N}}, \; with \; n = 0, 1, 2, \cdots N - 1. \tag{1.22}$$

Equation (1.22) is the definition of the DFT of N input samples taken at intervals of ΔT. The DFT is symmetric about $N/2$, the components beyond $N/2$ simply belong to negative frequency. Thus the DFT does not calculate frequency components beyond $N/(2T_0)$, which also happens to be the Nyquist limit to avoid aliasing errors.

Also note that any real function of time can be written as a sum of a real and an odd function. Consequently, by Properties 2 and 3 above any real function of time will have real parts of the DFT as even functions of frequency and the imaginary parts of the DFT will be odd functions of frequency.

1.5.1 DFT and Fourier series

The Fourier series coefficients of a periodic signal can be obtained from the DFT of its sampled data by dividing the DFT by N, the number of samples in the data window. Thus, the Fourier series for a function $x(t)$ can be expressed by the formula

$$x(t) = \sum_{k=-\infty}^{\infty} \alpha_k e^{\frac{j2\pi kt}{T}} = \sum_{k=-\infty}^{\infty} \left[\frac{1}{N} \sum_{n=0}^{N-1} x(k\Delta T) e^{\frac{-j2\pi kn}{N}} \right] e^{\frac{j2\pi kt}{T}}. \tag{1.23}$$

As there are only N components in the DFT, the summation on k in Eq. (1.23) is from $\{k = 0, \cdots, N-1\}$.

Example 1.3
Consider a periodic function $x(t) = 1 + \cos 2\pi f_0 t + \sin 2\pi f_0 t$. The function is already expressed in terms of its Fourier series, with $a_0 = 2$, $a_1 = 1$, and $b_1 = 1$. The signal is sampled 16 times in one period of the fundamental frequency. The sampled data, the DFT, and the DFT divided by 16 (N, the number of samples) is shown in Table 1.1.

Table 1.1 Sampled data and Fourier transform of the periodic function $t = 1 + \cos 2\pi f_0 t + \sin 2\pi f_0 t$

Sample no.	$x(t)$	Frequency	DFT	X = DFT/16
0	2.0000	0	16.0000	1.000
1	2.3066	f_0	$8.0000 + j8.0000$	$0.5000 + j0.5000$
2	2.4142	$2f_0$	$0.0000 - j0.0000$	$0.0000 + j0.0000$
3	2.3066	$3f_0$	$0.0000 - j0.0000$	$0.0000 + j0.0000$
4	2.0000	$4f_0$	$0.0000 + j0.0000$	$0.0000 + j0.0000$
5	1.5412	$5f_0$	$-0.0000 + j0.0000$	$0.0000 + j0.0000$
6	1.0000	$6f_0$	$0.0000 + j0.0000$	$0.0000 + j0.0000$
7	0.4588	$7f_0$	$0.0000 - j0.0000$	$0.0000 + j0.0000$
8	0.0000	–	-0.0000	$0.0000 + j0.0000$
9	-0.3066	$-7f_0$	$0.0000 + j0.0000$	$0.0000 + j0.0000$
10	-0.4142	$-6f_0$	$0.0000 - j0.0000$	$0.0000 + j0.0000$
11	-0.3066	$-5f_0$	$-0.0000 - j0.0000$	$0.0000 + j0.0000$
12	-0.0000	$-4f_0$	$0.0000 + j0.0000$	$0.0000 + j0.0000$
13	0.4588	$-3f_0$	$0.0000 + j0.0000$	$0.0000 + j0.0000$
14	1.0000	$-2f_0$	$0.0000 + j0.0000$	$0.0000 + j0.0000$
15	1.5412	$-f_0$	$8.0000 - j8.0000$	$0.5000 - j0.5000$

The last column contains the Fourier series coefficients. Note that the DC component a_0 appears in the 0th position, while the fundamental frequency component appears in the 2nd and 15th position. The cosine term being an even function produces real parts which are even functions of frequency (0.5 at $\pm f_0$), while the sine term is an odd function of time and produces odd functions of frequency ($\pm j0.5$ at $\pm f_0$). The coefficient a_1 is obtained by adding the real parts corresponding to f_0 and $-f_0$ in the (DFT/16) column, while the coefficient b_1 is obtained by subtracting the imaginary part of the $-f_0$ term from the imaginary part of the f_0 term:

$a_0 = 2X_0 = 2$
$a_1 = \text{Real}(X_1 + X_{N-1}) = 1$
$b_1 = \text{Imaginary}(X_1 - X_{N-1}) = 1$

From the above example it is clear that for real functions $x(t)$ the Fourier series coefficients of a periodic function can be obtained from the DFT of its sampled data by the following formulas:

$a_0 = 2.X_0$
$a_k = 2.\text{Real}(X_k)$
$b_k = 2.\text{Imaginary}(X_k)$ for $k = 1, 2, \cdots, N/2 - 1$.

1.5.2 DFT and phasor representation

A sinusoid $x(t)$ with frequency kf_0 with a Fourier series

$$x(t) = a_k \cos(2\pi kf_0 t) + b_k \sin(2k\pi f_0 t)$$

$$= \left\{ \sqrt{(a_k^2 + b_k^2)} \right\} \cos(2\pi kf_0 t + \phi) \quad \text{where } \phi = \arctan(\frac{-b_k}{a_k}) \qquad (1.24)$$

has a phasor representation (see Section 1.2)

$$X_k = \frac{1}{\sqrt{2}} \left\{ \sqrt{(a_k^2 + b_k^2)} \right\} e^{j\phi}, \qquad (1.25)$$

where the square-root of 2 in the denominator is to obtain the rms value of the sinusoid. The phasor in complex form becomes

$$X_k = \frac{1}{\sqrt{2}} (a_k - jb_k). \qquad (1.26)$$

Using the relationship of the Fourier series coefficients with the DFT, the phasor representation of the kth harmonic component is given by

$$X_k = \frac{1}{\sqrt{2}} \frac{2}{N} \sum_{n=0}^{N-1} x(n\Delta T) e^{-\frac{j2\pi kn}{N}}$$

$$= \frac{\sqrt{2}}{N} \sum_{n=0}^{N-1} x(n\Delta T) \left\{ \cos(\frac{2\pi kn}{N}) - j \sin(\frac{2\pi kn}{N}) \right\}.$$

(1.27)

Using the notation $x(n\Delta T) = x_n$, and $2\pi/N = \theta$ (θ is the sampling angle measured in terms of the period of the fundamental frequency component)

$$X_k = \frac{\sqrt{2}}{N} \sum_{n=0}^{N-1} x_n \left\{ \cos(kn\theta) - j \sin(kn\theta) \right\}.$$

(1.28)

If we define the cosine and sine sums as follows:

$$X_{kc} = \frac{\sqrt{2}}{N} \sum_{n=0}^{N-1} x_n \cos(kn\theta),$$

(1.29)

$$X_{ks} = \frac{\sqrt{2}}{N} \sum_{n=0}^{N-1} x_n \sin(kn\theta),$$

(1.30)

then the phasor X_k is given by

$$X_k = X_{kc} - jX_{ks}.$$

(1.31)

Equations (1.29) through (1.31) will be used to represent the phasor in most of the computations in the rest of our discussion.

Example 1.4
Consider a signal consisting of a DC component, and 60 Hz, 120 Hz, and 300 Hz components:

$x(t) = 0.5 + \cos(120\pi t + \pi/4) + 0.2 \cos(240\pi t + \pi/8) + 0.3 \cos(600\pi t)$.

Note that the signal is real, but not an even or odd function of time, and hence by Property 4 above, the real part of the Fourier transform will be even, and the imaginary part will be odd functions of frequency.

The signal is sampled at 1440 Hz, and the following 24 samples are obtained over a window of 16.66 ms, which corresponds to one period of the 60-Hz signal. There will be 24 frequency samples of the DFT. They are calculated and tabulated in Table 1.2.:

Table 1.2 Spectrum created by DFT

Sample no.	$x(k)$	Frequency	DFT	DFT/24
0	1.6919	0	$12.0000 + j0.0000$	$0.5000 + j0.0000$
1	1.1994	f_0	$8.4853 - j8.4853$	$0.3535 - j0.3535$
2	0.5251	$2f_0$	$2.2173 - j0.9184$	$0.0924 - j0.0383$
3	0.2113	$3f_0$	$0.0000 - j0.0000$	$0.0000 - j0.0000$
4	0.2325	$4f_0$	$0.0000 - j0.0000$	$0.0000 - j0.0000$
5	0.0915	$5f_0$	$3.6000 - j0.0000$	$0.1500 - j0.0000$
6	−0.3919	$6f_0$	$-0.0000 - j0.0000$	$-0.0000 - j0.0000$
7	−0.7776	$7f_0$	$0.0000 - j0.0000$	$0.0000 - j0.0000$
8	−0.6420	$8f_0$	$-0.0000 - j0.0000$	$-0.0000 - j0.0000$
9	−0.2113	$9f_0$	$-0.0000 - j0.0000$	$-0.0000 - j0.0000$
10	−0.0474	$10f_0$	$-0.0000 + j0.0000$	$-0.0000 + j0.0000$
11	−0.2454	$11f_0$	$0.0000 - j0.0000$	$0.0000 - j0.0000$
12	−0.3223	−	$0.0000 + j0.0000$	$0.0000 + j0.0000$
13	0.0441	$-11f_0$	$0.0000 + j0.0000$	$0.0000 + j0.0000$
14	0.5271	$-10f_0$	$-0.0000 - j0.0000$	$-0.0000 - j0.0000$
15	0.6356	$-9f_0$	$-0.0000 + j0.0000$	$-0.0000 + j0.0000$
16	0.4501	$-8f_0$	$-0.0000 + j0.0000$	$-0.0000 + j0.0000$
17	0.5119	$-7f_0$	$0.0000 + j0.0000$	$0.0000 + j0.0000$
18	1.0223	$-6f_0$	$-0.0000 + j0.0000$	$-0.0000 + j0.0000$
19	1.5341	$-5f_0$	$3.6000 + j0.0000$	$0.1500 + j0.0000$
20	1.5898	$-4f_0$	$0.0000 + j0.0000$	$0.0000 + j0.0000$
21	1.3644	$-3f_0$	$0.0000 + j0.0000$	$0.0000 + j0.0000$
22	1.3648	$-2f_0$	$2.2173 + j0.9184$	$0.0924 + j0.0383$
23	1.6420	$-f_0$	$8.4853 + j8.4853$	$0.3535 + j0.3535$

The Fourier series coefficients are

$a_0 = 1.0,$
$a_1 = 0.707,$
$b_1 = -0.707,$
$a_2 = 0.1848,$
$b_2 = -0.0766,$
$a_5 = 0.3,$
$b_5 = 0.000,$

leading to the Fourier series

$x(t) = 0.5 + 0.707\cos(120\pi t) - 0.707\sin(120\pi t) + 0.1848\cos(240\pi t) - 0.0766\sin(240\pi t) + 0.3\cos(600\pi t)$

which agrees with the expression for the input signal.

1.6 Leakage phenomena

The calculation of the DFT implies truncation of the sampled data outside the
data window. As shown in Figure 1.9 (c), the effect of sampling and windowing is to create a periodic function which replicates the samples of the original
function in repeating data windows. In general this new function has discontinuities at the window boundaries, and these discontinuities lead to a spurious
spectrum which is a continuous function of frequency. The side lobes of the
Fourier transform of the windowing function are superimposed on the spectrum of the original signal in the data window, and leads to errors in the Fourier transform calculated from the sampled data. This phenomenon is known
as the "leakage effect".

Example 1.5
Consider an input signal with a frequency 60.05 Hz being sampled at 1440
Hz, $x(t) = \cos(120.1\pi t)$ and $x_k = \cos(120.1 k\pi t/1440)$, $k = 0,1, \cdots,23$.

Table 1.3 Leakage effect in DFT calculations

Sample no.	$x(k)$	Frequency	DFT	DFT/24
0	1.0000	0	$0.0199 + j0.0000$	$0.0008 + j0.0000$
1	0.9659	f_0	$12.0048 - j0.0314$	$0.5002 - j0.0013$
2	0.8658	$2f_0$	$-0.0068 + j0.0000$	$-0.0003 + j0.0000$
3	0.7066	$3f_0$	$-0.0026 + j0.0000$	$-0.0001 + j0.0000$
4	0.4992	$4f_0$	$-0.0014 + j0.0000$	$-0.0001 + j0.0000$
5	0.2578	$5f_0$	$-0.0010 + j0.0000$	$-0.0000 + j0.0000$
6	-0.0013	$6f_0$	$-0.0007 + j0.0000$	$-0.0000 + j0.0000$
7	-0.2603	$7f_0$	$-0.0005 + j0.0000$	$-0.0000 + j0.0000$
8	-0.5015	$8f_0$	$-0.0005 + j0.0000$	$-0.0000 + j0.0000$
9	-0.7085	$9f_0$	$-0.0004 + j0.0000$	$-0.0000 + j0.0000$
10	-0.8671	$10f_0$	$-0.0004 + j0.0000$	$-0.0000 + j0.0000$
11	-0.9665	$11f_0$	$-0.0003 + j0.0000$	$-0.0000 + j0.0000$
12	-1.0000	–	$-0.0003 + j0.0000$	$-0.0000 - j0.0000$
13	-0.9652	$-11f_0$	$-0.0003 - j0.0000$	$-0.0000 - j0.0000$
14	-0.8645	$-10f_0$	$-0.0004 - j0.0000$	$-0.0000 - j0.0000$
15	-0.7048	$-9f_0$	$-0.0004 - j0.0000$	$-0.0000 - j0.0000$
16	-0.4970	$-8f_0$	$-0.0005 - j0.0000$	$-0.0000 - j0.0000$
17	-0.2552	$-7f_0$	$-0.0005 - j0.0000$	$-0.0000 - j0.0000$
18	0.0039	$-6f_0$	$-0.0007 - j0.0000$	$-0.0000 - j0.0000$
19	0.2628	$-5f_0$	$-0.0010 - j0.0000$	$-0.0000 - j0.0000$
20	0.5038	$-4f_0$	$-0.0014 - j0.0000$	$-0.0001 - j0.0000$
21	0.7103	$-3f_0$	$-0.0026 - j0.0000$	$-0.0001 - j0.0000$
22	0.8684	$-2f_0$	$-0.0068 - j0.0000$	$-0.0003 - j0.0000$
23	0.9672	$-f_0$	$12.0048 + j0.0314$	$0.5002 + j0.0013$

The Fourier series coefficient for the fundamental frequency is
$$a_1 = 2.\text{Real } X_1(f),$$
$$b_1 = 2.\text{Imaginary } X_1(f).$$
The phasor in polar coordinates is found to be $(1.0004/\sqrt{2})\angle 0.1499°$. The true value of the phasor as seen from the expression for $x(t)$ is of course $(1.0/\sqrt{2})\angle 0°$. The computation error is due to the leakage effect. In a later chapter we will consider the off-nominal frequency phasor estimation in greater detail and offer alternate methods of eliminating the small error introduced by the leakage effect.

Figure 1.5 shows the Fourier transform of a single square wave which is repeated in Figure 1.10 (a). Fourier transform of the square wave shown in Figure 1.5 (b) is repeated in Figure 1.10 (b). The Fourier transform of the square windowing function used in calculating the DFT has side-lobes as shown in Figure 1.10(b), which are responsible for the leakage effect. It is possible to use other types of windowing functions which produce side lobes which are smaller than those produced by the square wave. A popular windowing function which has this property is the Hanning function, given by

$$h(t) = 0.5 \, (1 + \cos \frac{2\pi t}{T_0}) \quad for \; -T_0/2 \le t \le T_0/2 . \tag{1.32}$$

The Fourier transform of the Hanning function is

$$H(f) = \frac{T_0/2}{(1/T_0^2 - f^2)} \frac{\sin(\pi f T_0)}{(\pi f T_0)} . \tag{1.33}$$

The Hanning function, and its Fourier transform, and the Fourier transform of the square window function are shown in Figure 1.10 (c) and (d).

It can be seen from Figure 1.10 that the Hanning windowing function has side lobes which are much smaller than those of the square windowing function. Thus using Hanning function in calculating a DFT leads to much smaller leakage effect and consequently the errors due to this effect are much reduced.

Another function known as the Hamming function is sometimes used as a windowing function. This function is quite similar to the Hanning function and is given by [12]:

$$h(t) = 0.54 + 0.46 \, (\cos \frac{2\pi t}{T_0}) \; for \; -T_0/2 \le t \le T_0/2. \tag{1.34}$$

Other windowing functions could be used to meet specific requirements regarding the leakage effect.

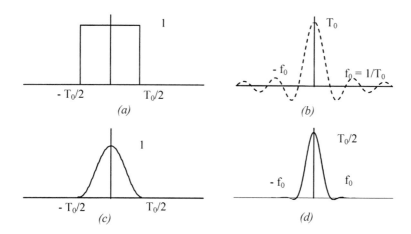

Fig. 1.10 (**a**) The rectangular window function and (**b**) its Fourier transform. The Hanning function (**c**), and its Fourier transform (**d**).

It is worth pointing out that in power system work, the principal contributor to the leakage effect are the off-nominal frequency input signals, when the sampling frequency is based upon the nominal power system frequency. For example, one may use a sampling frequency of 1440 Hz, which corresponds to 24 samples per period of the system nominal frequency of 60 Hz, while the actual power system frequency may be different from 60 Hz. Power system frequency never deviates from the nominal value by more than a few milli-Hertz. In such a case, the discontinuity at the window boundary is quite small, and the leakage effect even with the square windowing function is quite small. It is therefore common practice in power system work to use the square windowing function because of its simplicity.

References

1. Missout,G. and Girard, P., "Measurement of bus voltage angle between Montreal and Sept-Iles", IEEE Transactions on PAS. Vol. 99, No. 2, March/April 1980, pp 536–539.
2. Missout, G., Beland,J., and Bedard, G., "Dynamic measurement of the absolute voltage angle on long transmission Lines", IEEE Transactions on PAS. Vol. 100, No. 11, November 1981, pp 4428–4434.
3. Bonanomi,P., "Phase angle measurements with synchronized clocks – principles and applications", IEEE Transactions on PAS. Vol. 100, No. 11, November 1981, pp 5036–5043.

4. Phadke, A.G., Hlibka, T., and Ibrahim, M., "Fundamental basis for distance re-laying with symmetrical components", IEEE Transactions on PAS. Vol. 96, No. 2, March/April, 1977, pp 635–646.

5. Phadke, A.G., Thorp, J.S, and Adamiak,M.G., "A new measurement technique for tracking voltage phasors, local system frequency, and rate of change of fre-quency", IEEE Transactions on PAS. Vol. 102, No. 5, May 1983, pp 1025–1038.

6. There is great wealth of information about the GPS system available in various technical publications. A highly readable account for the layman is available at the web-site http://en.wikipedia.org/wiki/Gps. There the interested reader will also find links to other source material.

7. "Macrodyne Model 1690 PMU Disturbance Recorder", Macrodyne Inc. 4 Chel-sea Place, Clifton Park, NY, 12065.

8. "IEEE Standard for Synchrophasors for Power Systems", C37.118–2005, pp 56–57, IEEE 1344–1995. Sponsored by the Power System Relaying Committee of the Power Engineering Society, pp 56–57.

9. Papoulis, A., "The fourier integral and its applications", McGraw-Hill, New York, 1962.

10. Brigham, E.O., "The fast fourier transform", Prentice Hall, Englewood Cliffs, 1974.

11. Phadke, A.G. and Thorp, J.S., "Computer relaying for power systems", Research Studies Press, Reprinted August 1994.

12. Walker, J.S., "Fast fourier transforms", Second Edition, CRC Press, 1996.

Chapter 2 Phasor Estimation of Nominal Frequency Inputs

2.1 Phasors of nominal frequency signals

Consider a constant input signal $x(t)$ at the nominal frequency of the power system f_0, which is sampled at a sampling frequency Nf_0. The sampling angle θ is equal to $2\pi/N$, and the phasor estimation is performed using Eqs. (1.25–1.27).

$$x(t) = X_m \cos(2\pi f_0 t + \phi) \tag{2.1}$$

The N data samples of this input x_n: $\{n = 0,1,2,\cdots,N-1)$ are

$$x_n = X_m \cos(n\theta + \phi). \tag{2.2}$$

Since the principal interest in phasor measurements is to calculate the fundamental frequency component, we will set $k = 1$ in Eqs. (1.25 – 1.27) to produce the fundamental frequency phasor obtained from the sample set x_n. The superscript $(N-1)$ is used to identify the phasor as having the $(N-1)$st sample as the last sample used in the phasor estimation.

$$X_c^{N-1} = \frac{\sqrt{2}}{N} \sum_{n=0}^{N-1} x_n \cos(n\theta) = \frac{\sqrt{2}}{N} \sum_{n=0}^{N-1} X_m \cos(n\theta + \phi)\cos(n\theta)$$

$$= \frac{\sqrt{2}}{N} X_m \sum_{n=0}^{N-1} [\cos(\phi)\cos^2(n\theta) - \frac{1}{2}\sin(\phi)\sin(2n\theta)] = \frac{X_m}{\sqrt{2}}\cos(\phi) \tag{2.3}$$

It is to be noted that the summation of the $\sin(2n\theta)$ term over one period is identically equal to zero, and that the average of the $\cos^2(n\theta)$ term over a period is equal to 1/2.

The sine sum is calculated in a similar fashion:

A.G. Phadke, J.S. Thorp, *Synchronized Phasor Measurements and Their Applications*, DOI: 10.1007/978-0-387-76537-2_2, © Springer Science+Business Media, LLC 2008

$$X_s^{N-1} = \frac{\sqrt{2}}{N} \sum_{n=0}^{N-1} x_n \sin(n\theta) = \frac{\sqrt{2}}{N} \sum_{n=0}^{N-1} X_m \cos(n\theta + \phi) \sin(n\theta)$$

$$= \frac{\sqrt{2}}{N} X_m \sum_{n=0}^{N-1} [\frac{1}{2}\cos(\phi)\sin(2n\theta) - \sin(\phi)\sin^2(n\theta)] \qquad (2.4)$$

$$= -\frac{X_m}{\sqrt{2}} \sin(\phi)$$

The phasor X^{N-1} is given by

$$X^{N-1} = X_c^{N-1} - jX_s^{N-1} = \frac{X_m}{\sqrt{2}}[\cos(\phi) + j\sin(\phi)] = \frac{X_m}{\sqrt{2}}e^{j\phi} \qquad (2.5)$$

It is to be understood that Eq. (2.5) gives the fundamental frequency phasor estimate even though the subscript $k = 1$ has been dropped for the sake of simplicity. The result obtained in Eq. (2.5) conforms with the phasor definition given in Chapter 1, and the phase angle ϕ of the phasor is the angle between the time when the first sample is taken (corresponding to $n = 0$) and the peak of the input signal.

2.2 Formulas for updating phasors

2.2.1 Nonrecursive updates

Considering that the phasor calculation is a continuous process, it is neces-sary to consider algorithms which will update the phasor estimate as newer data samples are acquired. When the Nth sample is acquired after the pre-vious set of samples has led to the phasor estimate given by Eq. (2.5), the simplest procedure would be to repeat the calculations implied in Eqs. (2.3–2.5) for the new data window which begins at $n = 1$ and ends at $n = N$.

$$X^{N-1} = \frac{\sqrt{2}}{N} \sum_{n=0}^{N-1} x_n [(\cos(n\theta) - j\sin(n\theta)]$$

$$X^{N} = \frac{\sqrt{2}}{N} \sum_{n=0}^{N-1} x_{n+1} [(\cos(n\theta) - j\sin(n\theta)]$$

$$(2.6)$$

The two windows are shown in Figure 2.1. Phasor 1 is the result of phasor estimation over window 1, while phasor 2 is calculated with the data in window 2. The first sample in window 1 is lagging the peak of the sinusoid by an angle ϕ, while the first sample of window 2 ($n = 1$) lags the peak by an angle ($\phi + \theta$), θ being the angle between samples.

It should be clear from Figure 2.1 that in general the phasor obtained from a constant sinusoid of nominal power system frequency by this technique will have a constant magnitude and will rotate in the counterclockwise direction by angle θ as the data window advances by one sample. Since the phasor calculations are performed fresh for each window without using any data from the earlier estimates, this algorithm is known as a "nonrecursive algorithm". Nonrecursive algorithms are numerically stable, but are somewhat wasteful of computation effort as will be seen in the following.

Figure 2.2 is another view of the nonrecursive phasor estimation process. As newer samples are obtained, the table of sine and cosine multipliers is moved down to match the new data window. In this figure the multipliers are viewed as samples of unit-magnitude sine and cosine waves at the nominal power system frequency. The new data window has $N - 1$ samples in common with the old data window. In actual computation these are simply stored as tables of sine and cosine, which are used repeatedly on each window as needed.

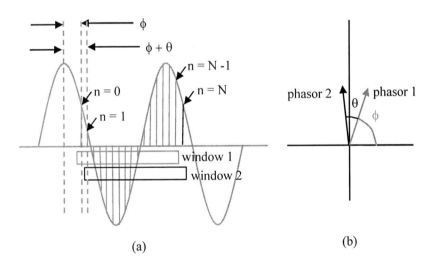

(a)   (b)

Fig. 2.1 Update of phasor estimates with N sample windows. Phasor 1 is calculated with samples $n = 0,\dots,N-1$, while phasor 2 is calculated with samples $n = 1,2,\dots,N$. θ is the angle between successive samples based on the period of the fundamental frequency.

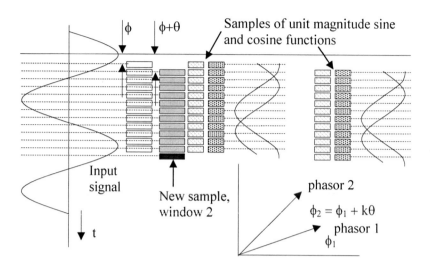

Fig. 2.2 Nonrecursive phasor estimation. There are 12 samples per cycle of the power frequency in this example. Fresh calculations are made for each new window as new samples are obtained. The phasor for a constant input signal rotates in the counterclockwise direction by the sampling angle, in this 30°.

2.2.2 Recursive updates

The formulas for calculating the $(N-1)$st and (N)th phasors by the nonrecursive algorithm are

$$X^{N-1} = \frac{\sqrt{2}}{N} \sum_{n=0}^{N-1} x_n e^{-jn\theta}$$

$$X^N = \frac{\sqrt{2}}{N} \sum_{n=0}^{N-1} x_{n+1} e^{-jn\theta}$$

(2.7)

The multipliers for a given sample are different in the two computations. For example, the multiplier for $(n = 2)$ sample in the first sum is $e^{-j2\theta}$ while the multiplier for the same sample in the second sum is $e^{-j\theta}$.

It should be noted that samples x_n: $\{n = 1,2,...,N-1\}$ are common to both windows. The second window has no x_0, so that it begins with x_1, and it ends with x_N, which did not exist in the first window. If one could arrange to keep the multipliers for the common samples the same in the two windows, one would save considerable computations in calculating X^N. If we multiply both sides of the second equation in (2.6) by $e^{-j\theta}$ we obtain the following result:

$$\hat{X}^N = e^{-j\theta}X^N = \frac{\sqrt{2}}{N}\sum_{n=0}^{N-1}x_{n+1}e^{-j(n+1)\theta}$$

$$= X^{N-1} + \frac{\sqrt{2}}{N}(x_N - x_0)e^{-j(0)\theta},$$

(2.8)

where use has been made of the fact that $e^{-j(0)\theta} = e^{-jN\theta}$, since N samples span exactly one period of the fundamental frequency. The phasor defined by Eq. (2.7) differs from the nonrecursive estimate by an angular retardation of θ. The advantage of using this alternative definition for the phasor from the new data window is that $(N-1)$ multiplications by the Fourier coefficients in the new window are the same as those used in the first window. Only a recursive update on the old phasor needs to be made to determine the value of the new phasor. This algorithm is known as the "recursive algorithm" for estimating phasors. In general, when the last sample in the data window is $(N+r)$, the recursive phasor estimate is given by

$$\hat{X}^{N+r} = e^{-j\theta}X^{N+r-1} + \frac{\sqrt{2}}{N}(x_{N+r} - x_r)e^{-jr\theta}$$

$$= \hat{X}^{N+r-1} + \frac{\sqrt{2}}{N}(x_{N+r} - x_r)e^{-jr\theta}.$$

(2.9)

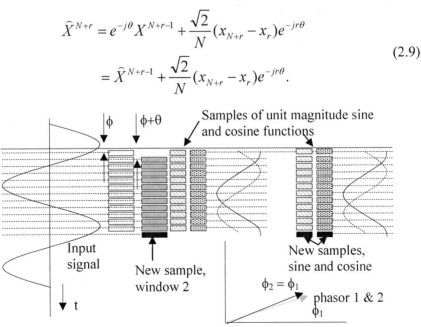

Fig. 2.3 Recursive phasor estimation. There are 12 samples per cycle of the power frequency in this example. Fresh calculations are made for each new window as new samples are obtained. New sine and cosine multipliers are used on the new sample. The phasor for a constant input signal remains stationary.

When the input signal is a constant sinusoid, x_{N+r} is the same as x_r, and the second term in Eq. (2.8) disappears. The phasor estimate with data from the new window is the same as the phasor estimate with data from the old window when the input signal is a constant sinusoid. In general, the recursive algorithm is numerically unstable. Consider the effect of an error in the estimate from one window – for example caused by a round-off error. This error is always present in all the phasor estimates from then on. This property of the recursive phasor algorithms must be kept in mind when practical implementation of these algorithms is performed [1]. Nevertheless, because of the great computational efficiency of the recursive algorithm, it is usually the algorithm of choice in many applications.

Unless stated otherwise explicitly, we will assume that only the recursive form of the phasor estimation algorithm is in use.

Example 2.1 Consider the 60-Hz signal $x(t) = 100 \cos(120\pi t + \pi/4)$ sampled at the rate of 12 samples per cycle, that is, at a sampling frequency of 720 Hz. The first 18 samples, and the nonrecursive and recursive phasor estimates obtained using Eqs. (2.6) and (2.8) beginning with sample no. 12 (at which time the first data window is completely filled) are shown in Table 2.1.

Table 2.1 Phasor estimates of sampled data

Sample no.	Sample x_n	Nonrecursive phasor estimate	Recursive phasor estimate
0	70.7107		
1	25.8819		
2	−25.8819		
3	−70.7107		
4	−96.5926		
5	−96.5926		
6	−70.7107		
7	−25.8819		
8	25.8819		
9	70.7107		
10	96.5926		
11	96.5926		
12	70.7107	70.701∠45°	70.701∠45°
13	25.8819	70.701∠75°	70.701∠45°
14	−25.8819	70.701∠105°	70.701∠45°
15	−70.7107	70.701∠135°	70.701∠45°
16	−96.5926	70.701∠165°	70.701∠45°
17	−96.5926	70.701∠195°	70.701∠45°

As expected, the nonrecursive phasor estimates produce a constant magnitude of $100/\sqrt{2}$ with an initial angle of $\pi/4$ (45°), and then for each successive estimate the angle increases by 30°.

2.3 Effect of signal noise and window length

The input signals are rarely free from noise. A spurious frequency component which is not a harmonic of the fundamental frequency signal may be
considered to be noise. One may also have induced electrical noise picked
up in the wiring of the input signal. Leakage effect caused by the windowing function has already been discussed in Chapter 1, and it too contributes
to an error in phasor estimation and should therefore be considered as a
type of noise in the input.

As an approximation, we will consider the noise in the input signal to be
a zero-mean, Gaussian noise process. This should be a good approximation
for the electrical noise picked up in the wiring and signal conditioning circuits. The other two sources of noise, namely nonharmonic frequency
components and leakage phenomena need further consideration. A phasor
measurement system may be placed in an arbitrarily selected substation
and will be exposed to input signals generated by the power system which
is likely to change states all the time. Each of the power system states may
lead to different nonharmonic frequencies and leakage effects, and the entire ensemble of conditions to which the phasor measurement system is
exposed may also be considered to be a pseudorandom Gaussian noise
process.

Consider a set of noisy measurement samples

$$x_n = X_m\cos(n\theta+\phi) +\varepsilon_n, \quad \{n = 0,1,2,\ldots,N-1\}, \tag{2.10}$$

where ε_n is a zero-mean Gaussian noise process with a variance of σ^2. If we
set $(X_m/\sqrt{2}) \cos(\phi) = X_r$, and $(X_m/\sqrt{2}) \sin(\phi) = X_i$, the phasor representing the
sinusoid is $X = X_r + jX_i$. We may pose the phasor estimation problem as
one of finding the unknown phasor estimate from the sampled data through
a set of N overdetermined equations:

$$\begin{bmatrix} x_0 \\ x_1 \\ x_2 \\ \cdot \\ x_{N-1} \end{bmatrix} = \sqrt{2}\begin{bmatrix} \cos(0) & -\sin(0) \\ \cos(\theta) & -\sin(\theta) \\ \cos(2\theta) & -\sin(2\theta) \\ \cdot & \cdot \\ \cos[(N-1)\theta] & -\sin[(N-1)\theta] \end{bmatrix}\begin{bmatrix} X_r \\ X_i \end{bmatrix} + \begin{bmatrix} \varepsilon_1 \\ \varepsilon_2 \\ \varepsilon_3 \\ \cdot \\ \varepsilon_{N-1} \end{bmatrix} \tag{2.11}$$

or, in matrix notation

$$[x] = [S][X]+[\varepsilon]. \tag{2.12}$$

Assuming that the covariance matrix $\mathbf{W}$ of the error vector is σ^2 multiplied by a unit matrix

$$[\mathbf{W}] = \sigma^2[\mathbf{1}] \qquad (2.13)$$

the weighted least-squares solution of Eq. (2.11) provides the estimate for the phasor

$$[\hat{\mathbf{X}}] = [\mathbf{S}^T \mathbf{W}^{-1} \mathbf{S}]^{-1} \mathbf{S}^T \mathbf{W}^{-1} [\mathbf{x}]. \qquad (2.14)$$

Using (2.13) for $\mathbf{W}$, and calculating $[\mathbf{S}^T \mathbf{S}]^{-1}$ for the S in Eq. (2.11)

$$[\hat{\mathbf{X}}] = [\mathbf{S}^T \mathbf{W}^{-1} \mathbf{S}]^{-1} [\mathbf{S}^T \mathbf{W}^{-1}][\mathbf{x}] = [\mathbf{S}^T \mathbf{S}]^{-1}[\mathbf{S}^T][\mathbf{x}] = \frac{1}{N}[\mathbf{S}^T][\mathbf{x}]. \qquad (2.15)$$

Since the noise is a zero-mean process, the estimate given by (2.15) is unbiased and the expected value of the estimate is equal to the true value of the phasor. If $\mathbf{X}$ is the true value of the phasor, the covariance matrix of the error in the phasor estimate is

$$E\left[[\hat{\mathbf{X}}\text{-}\mathbf{X}][\hat{\mathbf{X}}\text{-}\mathbf{X}]^T \right] = [\mathbf{S}^T \mathbf{W}^{-1} \mathbf{S}]^{-1}. \qquad (2.16)$$

Substituting for $[\mathbf{W}]$ from Eq. (2.13), the covariance of the error in phasor estimate is (σ^2/N). The standard deviations of error in real and imaginary parts of the phasor estimate are $(\sigma/\sqrt{N})$. We may thus conclude that higher sampling rates will produce improvement in phasor estimates in inverse proportion of the square root of the number of samples per cycle. Alternatively, if longer data windows are used (multiple of cycles), then once again the errors in phasor estimate go down as the square root of the number of cycles used. Thus, a four-cycle phasor estimate is twice as accurate as a one-cycle estimate in with noisy input.

Example 2.2 Consider a 60-Hz sinusoid $x(t) = 100 \cos(120\pi t + \pi/4) + \varepsilon(t)$ in a noisy environment, with the Gaussian noise ε having a zero mean and a standard deviation of 1. The source of noise could be electromagnetic interference, quantization errors, or harmonic and nonharmonic components in the input signal. If the phasing of the harmonic and nonharmonic signals is random, the noise model may be approximated by a zero-mean Gaussian characteristic.

The signal is sampled at six different sampling rates: 8, 16, 32, 64, 128, and 256 times per cycle. The signal samples are created with appropriately modeled noise input for 1000 cycles, and 1000 estimates of the phasor value are calculated. The standard deviations of the errors in the 1000 phasor estimates as well as its theoretical value ($\sigma/\sqrt{N}$) are given in the Table 2.2, and are also shown in Figure 2.4.

Table 2.2 Phasor estimation of a noisy signal

No. of samples per cycle(N)	Standard devia-tion of input noise	Standard deviation of phasor estimate error (volts)	$\sigma/\sqrt{N}$
8	1	0.3636	0.3536
16	1	0.2601	0.2500
32	1	0.1794	0.1768
64	1	0.1231	0.1250
128	1	0.0880	0.0884
256	1	0.0626	0.0625

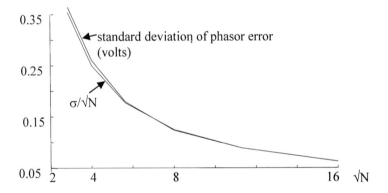

Fig. 2.4 The standard deviation of phasor error due to zero-mean Gaussian noise in the input. The result of 1000 phasor estimates at each sampling rate are shown by the solid line, and the theoretical value of the standard deviation ($\sigma/\sqrt{N}$) is shown by the dotted line.

These results show very good agreement with the expected results. As mentioned earlier, increasing the data window size at a fixed sampling rate, rather than the number of samples in the same data window, will produce similar results.

2.3.1 Errors in sampling times

Another possible source of error in input data samples is the error in timing of the sampling pulses. One possible source of errors is that the sampling clock is not precisely at a multiple of the power system frequency. This case will be dealt with in the next chapter, when we consider the phasor estimation problem at off-nominal frequencies.

   In this section we will consider the case where the sample times are corrupted by a Gaussian random noise with standard deviation varying by up to 10% of the sampling interval. Large errors of this type should not exist in modern measuring systems. Nevertheless, when sampling pulses are generated with the help of software clocks, it is possible to encounter random errors in sampling times. The following numerical example considers errors of this type. The errors are truncated at three times the standard deviation in order to eliminate impossibly large sampling clock errors.

Example 2.3 Consider a 60-Hz sinusoid $x(t) = 100 \cos(120\pi t + \pi/4)$ which is sampled at $t_n = n \, \Delta T + \varepsilon$, where the Gaussian noise ε has a zero mean and a standard deviation of $b \, \Delta T$, with the parameter b varying between 0.0 through 0.10. The signal is sampled at a sampling rate of 32 times per cycle. The signal samples are created with sampling time errors for 1000 cycles, and 1000 estimates of the phasor value are calculated. The standard deviations of the errors in the 1000 phasor estimates are given in the Table 2.3, and are also shown in Figure 2.5.

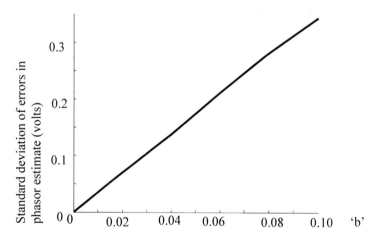

Fig. 2.5 The standard deviation of phasor error due to zero-mean Gaussian noise in sampling clock pulses. The result of 1000 phasor estimates at different values of '*b*' coefficients are shown.

Table 2.3 Errors in phasor estimation due to noisy inputs

Coefficient 'b' of input noise described above	Standard deviation of phasor estimate error (volts)
0.00	0.0000
0.02	0.0705
0.04	0.1373
0.06	0.2111
0.08	0.2809
0.10	0.3432

2.4 Phasor estimation with fractional-cycle data window

The weighted least-squares solution technique developed in Section 2.2 is a convenient vehicle for calculating phasors from fractional-cycle data windows. It should be remembered that fractional-cycle phasor estimates are necessary in developing high-speed relaying applications, and not particularly useful in wide-area phasor measurement applications whereby measurement times of a few cycles are acceptable. Nevertheless, it is instructive to include a discussion of fractional-cycle phasor estimation.

Consider the use of M samples of a sinusoid for estimating phasors, the sinusoid having been sampled at a sampling rate of N samples per cycle. $M < N$ produces a fractional-cycle phasor estimation algorithm.

As before, the input is a set of noisy measurement samples

$$x_n = X_m \cos(n\theta + \phi) + \varepsilon_n, \quad \{n = 0, 1, 2, \ldots, M-1\}, \tag{2.17}$$

where ε_n is a zero-mean Gaussian noise process with a variance of σ^2. The sampling angle θ is equal to $2\pi/N$.

$$\begin{bmatrix} x_0 \\ x_1 \\ x_2 \\ \cdot \\ x_{M-1} \end{bmatrix} = \sqrt{2} \begin{bmatrix} \cos(0) & -\sin(0) \\ \cos(\theta) & -\sin(\theta) \\ \cos(2\theta) & -\sin(2\theta) \\ \cdot & \cdot \\ \cos[(M-1)\theta] & -\sin[(M-1)\theta] \end{bmatrix} \begin{bmatrix} X_r \\ X_i \end{bmatrix} + \begin{bmatrix} \varepsilon_1 \\ \varepsilon_2 \\ \varepsilon_3 \\ \cdot \\ \varepsilon_{M-1} \end{bmatrix} \tag{2.18}$$

or, in matrix notation

$$[x] = [S][X] + [\varepsilon]. \tag{2.19}$$

As before, the weighted least-squares solution of Eq. (2.17) provides the estimate for the phasor

$$[\hat{X}] = [S^T W^{-1} S]^{-1} S^T W^{-1} [x] . \qquad (2.20)$$

Using (2.13) for W, and calculating $[S^T S]^{-1}$ for the S in Eq. (2.11):

$$[\hat{X}] = [S^T W^{-1} S]^{-1} [S^T W^{-1}][x] = [S^T S]^{-1} [S^T][x]. \qquad (2.21)$$

Unlike in the case of the full-cycle phasor estimation, $[S^T S]^{-1}$ is no longer a simple matrix:

$$[S^T S] = 2 \begin{bmatrix} \sum_{n=0}^{M-1} \cos^2(n\theta) & \sum_{n=0}^{M-1} \cos(n\theta)\sin(n\theta) \\ \sum_{n=0}^{M-1} \cos(n\theta)\sin(n\theta) & \sum_{n=0}^{M-1} \sin^2(n\theta) \end{bmatrix} . \qquad (2.22)$$

It can be shown that for a half-cycle estimation, with $M = N/2$, the least-squares solution is very similar to the DFT estimator.

2.5 Quality of phasor estimate and transient monitor

Phasor estimates obtained from a data window represent the fundamental frequency component of the input confined to the data window. When a fault occurs on the power system, there is a series of data windows which contain pre- and post-fault data. This is illustrated in Figure 2.6 for an assumed voltage waveform during a fault.

It should be clear that although a phasor estimate will be available for all data windows (including the ones that are shaded in Figure 2.6), only phasors which belong entirely to the pre- or post-fault periods are of interest. The phasors computed for the shaded windows of Figure 2.6 do not represent any meaningful system state, and a technique is needed to detect the occurrence of mixed states within a data window.

A technique known as "transient monitor" [2] provides a measure to indicate a "quality" of the estimate, and can also be used to detect the condition when a data window contains mixed-state waveforms. Consider the process of computing the data samples $(\hat{x}_n)$ in a window from the estimated phasor which has been estimated from a sample set (x_n):

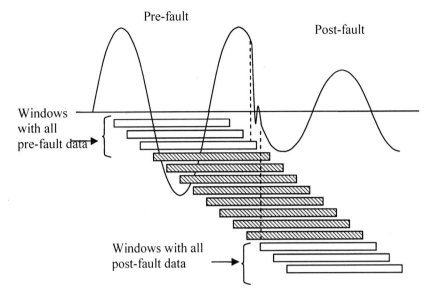

Fig. 2.6 Transition from pre-fault to post-fault waveforms. The shaded windows contain mixed waveform data.

$$[\hat{x}_n] = \sqrt{2} \begin{bmatrix} \cos(0) & -\sin(0) \\ \cos(\theta) & -\sin(\theta) \\ \cos(2\theta) & -\sin(2\theta) \\ \cdot & \cdot \\ \cos[(N-1)\theta] & -\sin[(N-1)\theta] \end{bmatrix} \begin{bmatrix} \hat{X}_r \\ \hat{X}_i \end{bmatrix}. \tag{2.23}$$

Substituting for the phasor estimate from Eq. (2.15)

$$[\hat{x}_n] = \sqrt{2} \begin{bmatrix} \cos(0) & -\sin(0) \\ \cos(\theta) & -\sin(\theta) \\ \cos(2\theta) & -\sin(2\theta) \\ \cdot & \cdot \\ \cos[(N-1)\theta] & -\sin[(N-1)\theta] \end{bmatrix}$$

$$\times \frac{\sqrt{2}}{N} \begin{bmatrix} \cos(0) & \cos(\theta) & \cos(2\theta) & \cdot & \cos[(N-1)\theta) \\ -\sin(0) & -\sin(\theta) & -\sin(2\theta) & \cdot & -\sin[(N-1)\theta) \end{bmatrix} [x_n]. \tag{2.24}$$

Multiplying the matrices and simplifying

$$[\hat{x}_n] = \frac{2}{N} \begin{bmatrix} 1 & \cos(\theta) & \cos(2\theta) & \cdot & \cos[(N-1)\theta] \\ \cos(\theta) & 1 & \cos(\theta) & \cdot & \cos(0) \\ \cos(2\theta) & \cos(\theta) & 1 & \cdot & \cos(\theta) \\ \cdot & \cdot & \cdot & 1 & \cdot \\ \cos[(N-1)\theta] & \cos(0) & \cos(\theta) & \cdot & 1 \end{bmatrix} [x_n] \tag{2.25}$$

where use has been made of the fact that $N\,\theta = 2\pi$. The difference between the input data and the recomputed sample data from the phasor estimate is the error of estimation $[t_n]$:

$$[t_n] = [x_n - \hat{x}_n]$$

$$= \begin{bmatrix} 1-\dfrac{2}{N} & -\dfrac{2}{N}\cos(\theta) & -\dfrac{2}{N}\cos(2\theta) & \cdot & -\dfrac{2}{N}\cos[(N-1)\theta] \\ -\dfrac{2}{N}\cos(\theta) & 1-\dfrac{2}{N} & -\dfrac{2}{N}\cos(3\theta) & \cdot & -\dfrac{2}{N}\cos(0) \\ -\dfrac{2}{N}\cos(2\theta) & -\dfrac{2}{N}\cos(3\theta) & 1-\dfrac{2}{N} & \cdot & -\dfrac{2}{N}\cos(\theta) \\ \cdot & \cdot & \cdot & 1-\dfrac{2}{N} & \cdot \\ -\dfrac{2}{N}\cos[(N-1)\theta] & -\dfrac{2}{N}\cos(0) & -\dfrac{2}{N}\cos(\theta) & \cdot & 1-\dfrac{2}{N} \end{bmatrix} [x_n] \tag{2.26}$$

If the input signal is a pure sinusoid at fundamental frequency, all entries of $[t_n]$ will be identically equal to zero. However, when the input signal is noisy or contains a composite window of two different sinusoids, $[t_n]$ is not zero, and one may use the sum (T_n) of the absolute values of its elements as a measure of the error of estimation:

$$T_n = \sum_{k=0}^{N-1} |t_k| \tag{2.27}$$

This sum has been referred as a "transient monitor", and can be used as a measure of the "quality" of the phasor estimate.

Example 2.4 Consider a composite 60-Hz voltage waveform samples described by

$$x_n = 100 \cos(n\theta + \pi/4), \text{ for } n = 0,1,2,\ldots,35$$
$$x_n = 50 \cos(n\theta + \pi/8), \text{ for } n = 36,37,38,\ldots,71$$

with a sampling rate of 24 samples per cycle; thus, $\theta = \pi/12$.
The data samples, recursive phasor estimates, and the function T_n are tabulated in Table 2.4:

Table 2.4 Transient monitor for a transient signal

Sample no.	Sample value	Phasor	T_n
1	70.7107	0	
2	50.0000	0	
23	96.5926	0	
24	86.6025	$50.0 + j50.0$	0.0000
36	−86.6025	$50.0 + j50.0$	0.0000
37	−46.1940	$48.5553 + j50.0$	51.4677
38	−39.6677	$47.9672 + j50.1576$	67.5480
39	−30.4381	$48.1998 + j50.0233$	68.1205
40	−19.1342	$48.9970 + j49.2261$	80.9808
41	−6.5263	$49.9518 + j47.5723$	129.1521
42	6.5263	$50.6149 + j45.0978$	195.2854
43	19.1342	$50.6149 + j42.0587$	267.1946
44	30.4381	$49.7583 + j38.8619$	331.2446
45	39.6677	$48.0811 + j35.9570$	382.1787
46	46.1940	$45.8392 + j33.7151$	413.6845
47	49.5722	$43.4397 + j32.3297$	430.7721
48	49.5722	$41.3320 + j31.7650$	436.9438
49	46.1940	$39.8874 + j31.7650$	432.8577
50	39.6677	$39.2993 + j31.9225$	426.1713
51	30.4381	$39.5318 + j31.7883$	431.3592
52	19.1342	$40.3290 + j30.9910$	433.7538
53	6.5263	$41.2839 + j29.3372$	424.5001
54	−6.5263	$41.9469 + j26.8628$	405.5234
55	−19.1342	$41.9469 + j23.8236$	367.7588
56	−30.4381	$41.0903 + j20.6269$	314.0214
57	−39.6677	$39.4132 + j17.7219$	243.3327
58	−46.1940	$37.1713 + j15.4800$	162.9157
59	−49.5722	$34.7718 + j14.0947$	77.7374
60	−49.5722	$32.6641 + j13.5299$	0.0000
72	49.5722	$32.6641 + j13.5299$	0.0000

Note that in the interest of saving space, several rows which do not show interesting transitions have been omitted. The phasor estimates and the transient monitor are plotted in Figure 2.7.

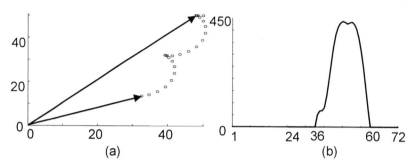

Fig. 2.7 Result of phasor estimation in a data stream with mixed input signals. (**a**) Phasor estimates. The transition from a solid phasor at $50 + j50$ to a new phasor of $32.6641 + j13.5299$ is shown by open circles. (**b**) The transient monitor T_n. Note that it is high during the transition from one phasor value to another. When the input signal is a pure sinusoid, the T_n becomes zero.

Note that the phasor estimate remains stationary at $(50 + j50)$ and $(32.6641 + j13.5299)$ while the input signal is $70.7\angle 45°$ and $50\angle 22.5°$, respectively, and the transition from one value to another takes 24 samples, the width of phasor estimation window. The transient monitor provides a good indication of the quality of the phasor estimate, it being high during the transition period when the phasor estimate is unreliable.

2.6 DC offset in input signals

Fault currents in a power system often have an exponentially decaying DC component, which is generally known as the DC offset. Occasionally, voltage waveforms may also have a DC offset due to capacitive voltage transformer transients. In both cases, the DC offsets decay to negligible values in a few cycles. If the phasor estimate is performed while a DC offset is present in a waveform, one is likely to get significant errors in phasor estimate while the DC offset is non-zero. The transient monitor described in Section 2.4 can be used to alert the user that the phasor estimate thus obtained is unreliable.

Since many phasor applications are dedicated to relatively slow phenomena, it is not essential that DC offset be handled in any special way; one only needs to be alert to the estimate quality indicated by the transient monitor. However, in computer relaying applications, powerful techniques have been developed to remove DC offsets before phasors are estimated, and in very specific applications of phasors which require very high speed of response, it may be necessary to employ algorithms which will remove the DC offset from the signals. This section provides a brief summary of the available techniques for this purpose. We will consider only the DC offset in current waveforms when a fault occurs. Similar techniques are applicable to the handling of DC offset in the voltage waveforms as well.

The earliest technique used in relays for removing the DC offset from fault currents is the one of using a "mimic" circuit in the secondary winding of a current transformer (CT)[3]. Figure 2.8 shows the primary fault circuit, and the CT secondary winding with a burden $(r + j\omega l)$ such that the ratio R/L is matched exactly by the burden ratio r/l. In this case, the DC offset in the current is not present in the voltage $e_2(t)$across the burden, and the burden voltage can be used as a signal which is proportional to the current and is free from the DC offset. The primary fault circuit and the CT secondary burden are both primarily inductive in nature, and hence the mimic circuit acts as a differentiator. Thus it has the property of amplifying any high-frequency noise that may be present in the current. However, it should be remembered that the primary fault current is itself produced by the R–L circuit, and thus has attenuated the high-frequency noise that may be present in the voltage signal. Thus, although the mimic circuit is a differentiator, the noise content of the voltage across it is similar to that in the primary voltage.

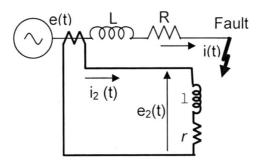

Fig. 2.8 Fault circuit and the mimic burden in CT secondary to eliminate the DC offset in fault current.

For computer relays, there is a least-squares solution technique available for eliminating the DC offset, which is free from the noise amplification properties of the mimic circuit. Consider a fault current $i(t)$ containing a DC offset given by

$$i(t) = A\cos(\omega t) + B\sin(\omega t) - C\varepsilon^{-t/T} \quad \text{for } t \geq 0$$
$$= A - C \quad \text{for } t = 0-.$$

(2.28)

This expression assumes that the current just before the occurrence of fault is $(A - C)$, and that the DC offset decays with a time constant T, which for the circuit of Figure 2.8 is equal to L/R seconds.

Consider a sample set of M data points obtained from this current waveform

$$i_n = A \cos(n\theta) + B \sin(n\theta) - Cr^n, \text{ for } \{n = 0,1,2,\ldots,M-1\}, \quad (2.29)$$

where θ is the sampling angle equal to $2\pi/N$, N being the number of samples per cycle of the nominal frequency, and r is the decrement factor for the decaying DC component in one sample time:

$$r = \varepsilon^{-\Delta T/T}. \quad (2.30)$$

If we now assume that the decrement factor r is known, the only unknowns in Eq. (2.30) are A, B, and C. Taking the overdetermined set of M data points,

$$\begin{bmatrix} i_0 \\ i_1 \\ \vdots \\ i_{M-1} \end{bmatrix} = \begin{bmatrix} 1 & 0 & -1 \\ \cos\theta & \sin\theta & -r \\ \vdots & \vdots & \vdots \\ \cos(M-1)\theta & \sin(M-1)\theta & -r^{M-1} \end{bmatrix} \begin{bmatrix} A \\ B \\ C \end{bmatrix}. \quad (2.31)$$

As usual, the above equation can be solved for A, B, and C, and then by adding (Cr^{-n}) to each sample of the current the DC offset can be removed from the waveform.

It is possible that the value of the time constant T is not known exactly and an approximate value must be used for "r". The algorithm is tolerant of reasonable errors in the value of "r", as is seen by the following numerical example.

Example 2.5 Consider a fault current waveform with full DC offset given by $i(t) = 100 \cos(120\pi t) + 100 \sin(120\pi t) - 100 \, e^{-t/0.05}$. The current is zero before the occurrence of the fault. The DC offset decay time constant is 50 ms.

The true value of the phasor must be $(100/\sqrt{2})(1 - j1)$. The DC offset is removed by applying Eq. (2.31) for a window of one cycle at a time. It is assumed that an error of $\pm 10\%$ is made in the decrement factor "r" used in Eq. (2.31). The resulting error in ($\sqrt{2}\times$phasor) is shown in Figure 2.9.

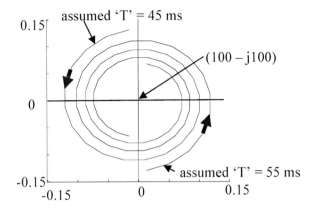

Fig. 2.9 Errors in phasor estimate caused by errors in time constant estimate. Two cycles worth of data is shown. The solid arrows show the direction of increasing time.

It can be seen from Figure 2.9 that the errors of estimation of phasors are less than 0.2% even though the time constant errors are of the order of 10%.

The least-squares solution described above is sometimes described as a "digital mimic" procedure. However, it must be pointed out that this process is not a differentiator, and consequently there is no amplification of noise in the current signal in this process.

2.7 Non-DFT estimators

A number of papers dealing with the problem of computing phasors from sampled data have been published over the last several years. Several papers consider variations on the Fourier transform method, with special emphasis on the problem of dealing with off-nominal frequency signals. We will consider such signals in the next chapter, where the performance of the fixed-frequency Fourier transform on off-nominal frequency input signals will be discussed. Among the variations of the basic Fourier technique, least-squares methods, Kalman filter methods, and Prony methods have been discussed. As the main thrust of these variations is to deal with off-nominal frequency signals, we will defer their discussion to a later chapter.

References

1. Phadke, A.G. and Thorp, J.S., "Computer relaying for power systems", Research Studies Press Ltd., John Wiley & Sons, Inc., 1994, pp 127–129.
2. Phadke, A.G. and Thorp, J.S., "Computer relaying for power systems", Research Studies Press Ltd., John Wiley & Sons, Inc., 1994, pp 151–152.
3. Mason, C.R.,"Art and science of protective relaying", John Wiley & Sons, 1956.

Chapter 3 Phasor Estimation at Off-Nominal Frequency Inputs

3.1 Types of frequency excursions found in power systems

Phasors are a steady-state concept. In reality, a power system is never in a steady state. Voltage and current signals have constantly changing fundamental frequency (albeit in a relatively narrow range around the nominal frequency) due to changes in load and generation imbalances and due to the interactions between real power demand on the network, inertias of large generators, and the operation of automatic speed controls with which most generators are equipped. In addition, when faults and other switching events take place, there are very rapid changes in voltage and current waveforms, and depending upon the definition of frequency one would have to accept that power system waveforms under these conditions contain a very wide band of frequencies ranging from *0 hertz* to hundreds of kilohertz. The consideration of various interpretations of frequency which are of interest in power system engineering will be considered in Chapters 4 and 6.

In this chapter we will focus our attention on the changes in power system frequency due to responses to load generation imbalances and when the power system is in a quasi-steady state and is operating with a frequency which may be different from its nominal value. It will be assumed that power system voltages and currents are balanced, and the frequency changes are only due to speed changes of the rotors of power system generators. As these speed changes are slow (as compared to the nominal power system frequency), one may consider the progress of such speed changes as a sequence of quasi-steady states when the waveforms are observed over a small window – for example, over one period of the power frequency.

Most integrated power systems operate in a relatively narrow band of frequency, within 0.5 Hz from its nominal value. Under exceptional cir-

A.G. Phadke, J.S. Thorp, *Synchronized Phasor Measurements and Their Applications*, DOI: 10.1007/978-0-387-76537-2_3, © Springer Science+Business Media, LLC 2008

cumstances – for example, when small islands of generators and load are isolated from the rest of the network – the frequency excursions may be as large as ±10 Hz. However, the power system operation at such extreme excursions are usually controlled and brought back to normal values by available control actions. Where the islands are primarily powered by hydroelectric generators, the system may operate at large frequency deviations for extended periods.

3.2 DFT estimate at off-nominal frequency with a nominal frequency clock

It is assumed that the sampling clock is a fixed-frequency clock with sampling rates which are multiples of the nominal power system frequency. Recursive phasor calculation formulas were developed in Chapter 2 (Eqs. 2.8 and 2.9) and are used to consider phasor estimation when the power system frequency differs from the nominal.

Equation (2.9) is reproduced here for ready reference as Eq. (3.1):

$$\widehat{X}^{N+r} = e^{-j\theta} X^{N+r-1} + \frac{\sqrt{2}}{N}(x_{N+r} - x_r)e^{-jr\theta}$$
$$= \widehat{X}^{N+r-1} + \frac{\sqrt{2}}{N}(x_{N+r} - x_r)e^{-jr\theta}.$$

(3.1)

It should be clear that if the input signal is a constant sinusoid of nominal power system frequency $x_{N+r} = x_r$, and consequently Eq. (3.1) confirms that the resulting phasor would also remain constant [1].

3.2.1 Input signal at off-nominal frequency

Now assume that the input signal is at a frequency

$$\omega = \omega_0 + \Delta\omega,$$

(3.2)

where ω_0 is the nominal power system frequency. For a 60-Hz system, ω_0 is 120π radians per second.

The input signal is once again assumed to be

$$x(t) = X_m \cos(\omega t + \phi).$$

(3.3)

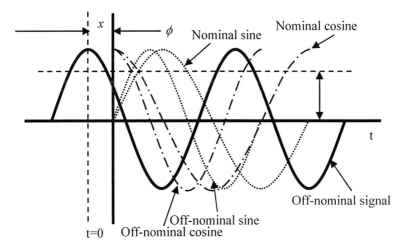

Fig. 3.1 Phasor calculation at off-nominal frequency signals with sampling clock synchronized with the nominal power system frequency.

The corresponding phasor representation is $(X_m/\sqrt{2})(\varepsilon^{j\phi})$, which is the same formula as given in Eq. (1.2). *In fact the definition of a phasor representation of a sinusoid is independent of the frequency of the signal.* We would get this phasor with the phasor computation formulas given above if the Fourier coefficients were also from sines and cosines of the same period. However, when fixed-frequency clocks related to the nominal power system frequency are used, the resulting phasors would be different from the true value given above. These considerations are illustrated in Figure 3.1.

Figure 3.1 shows the off-nominal frequency being much higher than the nominal frequency. As mentioned before, in normal power systems the deviation in frequency would certainly be much smaller. However, the figure illustrates the concept involved. If the phasor calculation is performed with off-nominal sine and cosine waves, the result will be the correct phasor for the given sinusoid. However, using the nominal frequency sines and cosines for phasor calculation will introduce an error in phasor estimation. Clearly the error made in phasor estimation will depend upon the difference between the nominal and actual frequency.

The input signal can be expressed as follows:

$$x(t) = X_m \cos(\omega t + \phi) = \sqrt{2}\ Re\ [(X_m/\sqrt{2})(\varepsilon^{j\phi})(\varepsilon^{j\omega t})]$$
$$= \sqrt{2}\ Re\ \{X\varepsilon^{j\omega t}\},$$

(3.4)

where X is the correct value of the phasor at the off-nominal frequency and the function "Re" is the real value function. Expressing the real value as the average of a complex number and its complex conjugate

$$x(t) = (\sqrt{2}/2)\{X\, \varepsilon^{j\omega t} + X^*\, \varepsilon^{-j\omega t}\}. \tag{3.5}$$

The kth sample of the signal represented by Eq. 3.5 is given by

$$x_k = (1/\sqrt{2})\{X\, \varepsilon^{j\omega k \Delta t} + X^*\, \varepsilon^{-j\omega k \Delta t}\} \tag{3.6}$$

The phasor representation of $x(t)$, that is, X' (which is different from X unless the system frequency is equal to the nominal value ω_0) – is calculated using Eq. (3.1) with x_r as the first sample. Note that Eq. (3.1) uses sine and cosine terms at the nominal power system frequency ω_0. Thus X_r' is given by

$$
\begin{aligned}
X_r' &= \frac{\sqrt{2}}{N} \sum_{k=r}^{r+N-1} x_k \varepsilon^{-jk\omega_0 \Delta t} \\
&= \frac{1}{N} \sum_{k=r}^{r+N-1} \{X \varepsilon^{jk\omega \Delta t} + X^* \varepsilon^{-jk\omega \Delta t}\} \varepsilon^{-jk\omega_0 \Delta t}.
\end{aligned}
\tag{3.7}
$$

Making use of the identity

$$
\begin{aligned}
\varepsilon^{jx} - 1 &= \varepsilon^{jx/2}(\varepsilon^{jx/2} - \varepsilon^{-jx/2}) \\
&= 2j\varepsilon^{jx/2} \sin(x/2)
\end{aligned}
$$

the two summations in Eq. (4.9), which are geometric series, can be expressed in closed form as [2]

$$
\begin{aligned}
X_r' &= X\varepsilon^{jr(\omega-\omega_0)\Delta t} \left\{ \frac{\sin \dfrac{N(\omega-\omega_0)\Delta t}{2}}{N\sin \dfrac{(\omega-\omega_0)\Delta t}{2}} \right\} \varepsilon^{j(N-1)\frac{(\omega-\omega_0)\Delta t}{2}} \\
&\quad + X^*\varepsilon^{-jr(\omega+\omega_0)\Delta t} \left\{ \frac{\sin \dfrac{N(\omega+\omega_0)\Delta t}{2}}{N\sin \dfrac{(\omega+\omega_0)\Delta t}{2}} \right\} \varepsilon^{-j(N-1)\frac{(\omega+\omega_0)\Delta t}{2}}
\end{aligned}
\tag{3.8}
$$

or

$$X_r^{'} = PX\varepsilon^{jr(\omega-\omega_0)\Delta t} + QX^*\varepsilon^{-jr(\omega+\omega_0)\Delta t}, \qquad (3.9)$$

where P and Q are coefficients in Eq. (3.9) which are independent of "r":

$$P = \left\{ \frac{\sin\dfrac{N(\omega-\omega_0)\Delta t}{2}}{N\sin\dfrac{(\omega-\omega_0)\Delta t}{2}} \right\}\varepsilon^{j(N-1)\frac{(\omega-\omega_0)\Delta t}{2}}, \qquad (3.10)$$

$$Q = \left\{ \frac{\sin\dfrac{N(\omega+\omega_0)\Delta t}{2}}{N\sin\dfrac{(\omega+\omega_0)\Delta t}{2}} \right\}\varepsilon^{-j(N-1)\frac{(\omega+\omega_0)\Delta t}{2}}. \qquad (3.11)$$

3.2.1.1 A qualitative graphical representation

The phasor estimate at off-nominal frequency is given by Eq. (3.9). It should be noted that for all practical power system frequencies, $\omega - \omega_0$ is likely to be very small, and hence ($\omega + \omega_0 = 2\omega_0 + \Delta\omega$) is very nearly equal to $2\omega_0$. A qualitative representation of Eq. (3.9) is shown in Figure 3.2.

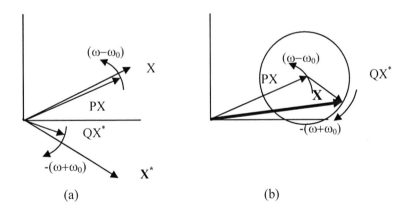

(a)   (b)

Fig. 3.2 Qualitative illustration of phasor estimates at off-nominal frequency.

In Figure 3.2, the phasors X and X^* are attenuated by complex gains P and Q as shown in Figure 3.2(a). PX rotates in the anticlockwise direction at an angular speed of $(\omega - \omega_0) = \Delta\omega$. The phasor QX^* rotates in the clockwise direction at a speed $(\omega + \omega_0)$, which is approximately equal to $2\omega_0$. Figure 3.2(b) shows the resultant phasor, which is made of the two components. The resultant phasor thus has a magnitude and phase angle variation at a frequency $2\omega_0$ (approximately) superimposed on a monotonically rotating component at $\Delta\omega$. The qualitative variation of the magnitude and phase angle of the estimate of an off-nominal input signal is shown in Figure 3.3.

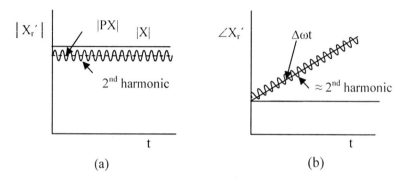

Fig. 3.3 Magnitude and angle variation with time of phasor estimate of an off-nominal signal.

Note that in Figure 3.3 the effect of the QX_Q term has been exaggerated in order to illustrate the behavior of the estimate. As will be seen in the next chapter, the actual effect is quite small when practical frequency excursions are considered.

The constants P and Q are complex numbers, and their values depend upon the deviation between the nominal frequency and the actual signal frequency. This dependence is illustrated in Figures 3.4 and 3.5 for a nominal frequency of 60 Hz, a frequency deviation in the range of ±5 Hz, and a sampling rate of 24 samples per cycle.

Note that the maximum attenuation occurs at a deviation of 5 Hz from nominal frequency, being around about 98.8%. For a 2-Hz deviation, the attenuation is at 99.8%, which for practical cases can be completely disregarded. The phase angle error corresponds to about 3 degrees per Hz deviation, varying linearly in the ±5-Hz range. Remembering that the factor P affects the principal term of the quantity being measured, the effect of this factor can often be neglected. For the sake of completeness, the data plotted in Figure 3.4 is also provided in Table 3.1.

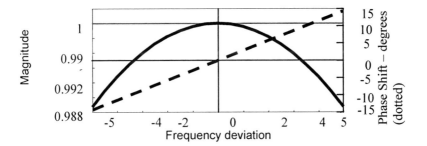

Fig. 3.4 The factor P as a function of frequency deviation.

Table 3.1 Magnitude and phase angle of P

Δf	$|P|$	$\angle P$ (degrees)
−5	0.9886	−14.37
−4.5	0.9908	−12.94
−4	0.9927	−11.5
−3.5	0.9944	−10.06
−3	0.9959	−8.62
−2.5	0.9972	−7.19
−2	0.9982	−5.75
−1.5	0.9990	−4.31
−1	0.9995	−2.87
−0.5	0.9999	−1.44
0	1.0000	0
0.5	0.9999	1.44
1	0.9995	2.87
1.5	0.9990	4.31
2	0.9982	5.75
2.5	0.9972	7.19
3	0.9959	8.62
3.5	0.9944	10.06
4	0.9927	11.5
4.5	0.9908	12.94
5	0.9886	14.37

The effect of sampling rate on the attenuation and phase shift is relatively minor. For example, for a +2.0-Hz deviation, by varying the sampling rate from 12 to 120 samples per cycle the effect on P is as shown in Table 3.2.

Table 3.2 Effect of the sampling rate on P for a frequency of 62 Hz

Sampling rate	$|P|$	$\angle P$ (degrees)
12	0.9982	5.5
24	0.9982	5.75
36	0.9982	5.83
48	0.9982	5.87
60	0.9982	5.9
72	0.9982	5.92
84	0.9982	5.93
96	0.9982	5.94
108	0.9982	5.94
120	0.9982	5.95

As can be seen from Table 3.2 the sampling rate affects the attenuation and the phase shift only slightly.

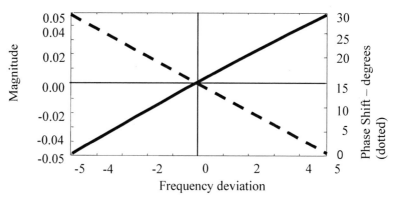

Fig. 3.5 The factor Q as a function of frequency deviation.

The effect of frequency deviation on the magnitude and phase angle of the attenuation factor Q is shown in Figure 3.5 for frequency excursions in the range of ±5 Hz.

Note that at the nominal frequency the magnitude of Q is 0. It increases almost linearly as a function of the frequency deviation, being about 0.008 per unit per Hz. Note also that at negative frequency deviations, the multiplier is also negative, so in that sense what is plotted is not the absolute value of Q. The phase angle of Q is 15 degrees at the nominal frequency, and the phase angle varies linearly with respect to frequency deviation. For the sake of completeness, the data plotted in Figure 3.5 is also provided in Table 3.3.

Table 3.3 Magnitude and phase angle of Q

Δf	$|Q|$	$\angle Q$ (degrees)
−5	−0.0434	29.37
−4.5	−0.0390	27.94
−4	−0.0346	26.5
−3.5	−0.0302	25.06
−3	−0.0258	23.62
−2.5	−0.0215	22.19
−2	−0.0171	20.75
−1.5	−0.0128	19.31
−1	−0.0085	17.87
−0.5	−0.0042	16.44
0	0	15.
0.5	0.0042	13.56
1	0.0084	12.12
1.5	0.0125	10.69
2	0.0166	9.25
2.5	0.0206	7.81
3	0.0246	6.37
3.5	0.0285	4.94
4	0.0324	3.5
4.5	0.0363	2.06
5	0.0400	0.62

The effect of sampling rate on the attenuation and phase shift of Q is also relatively minor. For example, for a +2.0-Hz deviation, by varying the sampling rate from 12 to 120 samples per cycle the effect on Q is as shown in Table 3.4.

Table 3.4 Effect of the sampling rate on Q for a frequency of 62 Hz

Sampling rate	$|Q|$	$\angle Q$ (degrees)
12	0.0172	24.5
24	0.0166	9.25
36	0.0164	4.17
48	0.0164	1.62
60	0.0164	0.1
72	0.0164	−0.92
84	0.0164	−1.64
96	0.0164	−2.19
108	0.0164	−2.61
120	0.0164	−2.95

As can be seen, the sampling rate does not affect the attenuation, but does affect the phase shift significantly.

Example 3.1 Numerical example: single-phase off-nominal frequency signal.

As an illustration of the application of the algorithms derived above, consider an example of a sinusoid having an rms value of 100 at a frequency of 60.5 Hz. Let the assumed phase angle of the phasor be $\pi/4$, so that the correct phasor representation of this signal is $X = 100e^{j\pi/4}$. Assume that this input signal is sampled at a frequency of 24 times the nominal frequency, or at 1440 Hz for a 60-Hz system. From Tables 3.1 and 3.3, the coefficients P and Q corresponding to this case ($\Delta f = +0.5$ Hz) are $P = 0.9999@1.44°$ and $0.0042@13.56°$, respectively.

The input signal is sampled at 1440 Hz, and 200 samples are processed by the Fourier formula [Eq. (3.1)]. Table 3.5 lists first 25 samples of the signal:

Table 3.5 First 25 samples of the 60.5-Hz signal $100e^{j\pi/4}$

Sample no.	x_k	Sample no.	x_k
1	100.0000	14	−67.2091
2	70.4433	15	−32.4138
3	36.0062	16	4.6272
4	−0.9256	17	41.3476
5	−37.7932	18	75.2034
6	−72.0424	19	103.8489
7	−101.3004	20	125.2995
8	−123.5400	21	138.0691
9	−137.2205	22	141.2730
10	−141.3941	23	134.6891
11	−135.7716	24	118.7737
12	−120.7424	25	94.6294
13	−97.3480		

The estimated phasor magnitude and angle calculated by the recursion formula for 160 samples of the input signal are shown in Figures 3.6 and 3.7. Note that the second harmonic ripple anticipated in the theory is evident in both figures. The amplitude of the ripple in the magnitude is about 0.42 (zero-to-peak of the variation in Figure 3.6), which is identical to the predicted value for $|Q|$ of 0.0042 for a frequency deviation of 0.5 Hz as seen from Table 3.3.

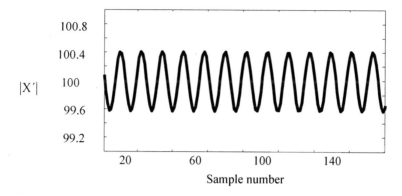

Fig. 3.6 Magnitude of phasor estimate of a signal at 60.5 Hz.

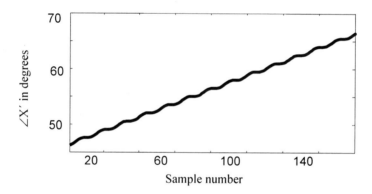

Fig. 3.7 Angle of phasor estimate of a signal at 60.5 Hz.

The estimated angle shown in Figure 3.7 contains an average slope corresponding to $\Delta\omega t$. Here also the second harmonic ripple in the angle estimate is evident.

3.3 Post processing for off-nominal frequency estimates

3.3.1 A simple averaging digital filter for $2f_0$

A very effective filter to correct for the errors introduced by the factor Q is to average three successive values of the estimate such that their relative phase angles are 60° and 120° at the nominal fundamental frequency, which would correspond to 120° and 240° for the second harmonic. The result of applying such a filter to the data in Figures 3.6 and 3.7 are shown in Figures 3.8 and 3.9, respectively. As can be seen, the second harmonic variation has been practically eliminated, and the remaining errors in the estimation are negligible.

Note that even after the single-phase signals were filtered with the three-point algorithm, there was a small amount of residual second harmonic ripple. This is to be expected, as the ripple in the single-phase estimation process is at $2\omega_0 + \Delta\omega$ rather than at $2\omega_0$, hence the three-point averaging does not eliminate the ripple exactly.

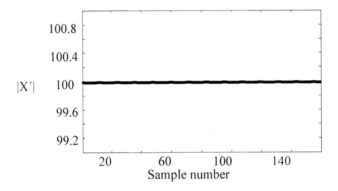

Fig. 3.8 Magnitude of phasor estimate of a signal at 60.5 Hz using three-point averaging.

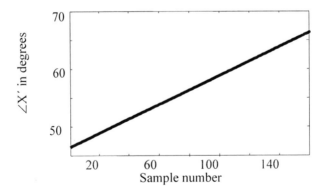

Fig. 3.9 Angle of phasor estimate of a signal at 60.5 Hz using three-point averaging.

3.3.2 A resampling filter

Another very effective filter is the resampling filter. Using the Fourier method to calculate the phasor using Eq. (3.9), the signal frequency is estimated by taking the derivative of the phasor angle (see Chapter 4). With this estimated frequency, the samples of the original signal at the estimated frequency are calculated using an interpolation formula so that the new sampling rate corresponds to the estimated frequency. Assuming that the input signal is a sinusoid, the interpolation formula can be derived as shown below.

Consider the input signal at frequency ω and a sampling clock corresponding to a frequency N times the nominal power system frequency ω_0. Using the notation of section 3.2, the sampling interval is $2\pi/N$ radians of the nominal frequency. Consider the samples corresponding to the sampling index "k" and "$k+1$" as shown in Figure 3.10 obtained at a sampling rate of N samples per cycle at the nominal frequency ω_0. It is required that sample number "m" corresponding to the sampling rate of N samples per cycle of the "estimated frequency ω" be calculated using an interpolation formula.

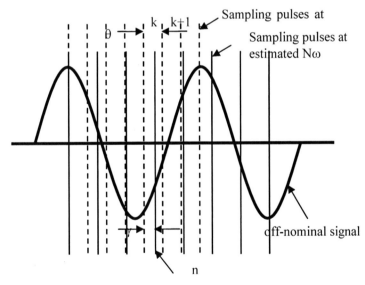

Fig. 3.10 Resampling process applied to an off-nominal signal.

Assuming that the input signal is given by

$$x(t) = X_m \cos(\omega t + \phi) \tag{3.12}$$

the samples corresponding to "k" and "$k+1$" are given by

$$x_k = X_m \cos[k\theta + \phi], \; x_{k+1} = X_m \cos[(k+1)\theta + \phi]. \tag{3.13}$$

It is required that the sample x_n corresponding to a sampling pulse generated by the sampling clock corresponding to the frequency ω is

$$x_n = X_m \cos[k\theta + \gamma + \phi], \tag{3.14}$$

where the angles θ and γ are expressed based on the estimated frequency ω. Using trigonometric identities, it can be shown that

$$x_n = x_k \{\sin(\theta - \gamma)\}/\sin\theta + x_{k+1}\{\sin\gamma\}/\sin\theta. \tag{3.15}$$

Phasor estimation is then performed using the resampled data. These resampled data phasors have very little errors of estimation.

Example 3.2 Phasor estimation with resampled data.
Consider the data in Table 3.5, which is obtained from a signal at 60.5 Hz. The data is taken at 60×24 samples per second. The resampled data obtained for samples obtained by sinusoidal interpolation as in Eq. (3.15) are given in Table 3.6.

Table 3.6 First 25 samples of the resampled 60.5-Hz signal $100e^{j\pi/4}$

Sample no.	x_k	γ	Resampled data	Sample no.	x_k	γ	Resampled data
1	100.0000		100.0000	14	−67.2091	0.2356	−70.7107
2	70.4433	0.2618	70.7107	15	−32.4138	0.2334	−36.6025
3	36.0062	0.2596	36.6025	16	4.6272	0.2313	−0.0000
4	−0.9256	0.2574	0.0000	17	41.3476	0.2291	36.6025
5	−37.7932	0.2553	−36.6025	18	75.2034	0.2269	70.7107
6	−72.0424	0.2531	−70.7107	19	103.8489	0.2247	100.0000
7	−101.3004	0.2509	−100.0000	20	125.2995	0.2225	122.4745
8	−123.5400	0.2487	−122.4745	21	138.0691	0.2203	136.6025
9	−137.2205	0.2465	−136.6025	22	141.2730	0.2182	141.4214
10	−141.3941	0.2443	−141.4214	23	134.6891	0.2160	136.6025
11	−135.7716	0.2422	−136.6025	24	118.7737	0.2138	122.4745
12	−120.7424	0.2400	−122.4745	25	94.6294	0.2116	100.0000
13	−97.3480	0.2378	−100.0000				

3.4 Phasor estimates of pure positive-sequence signals

3.4.1 Symmetrical components

The symmetrical components of three-phase voltages and currents are defined by the formula

$$
\begin{bmatrix} X_0 \\ X_1 \\ X_2 \end{bmatrix} = \frac{1}{3} \begin{bmatrix} 1 & 1 & 1 \\ 1 & \alpha & \alpha^2 \\ 1 & \alpha^2 & \alpha \end{bmatrix} \begin{bmatrix} X_a \\ X_b \\ X_c \end{bmatrix},
\tag{3.16}
$$

where the phase quantity phasors are used to calculate the symmetrical components. Note that we are considering phasors calculated according to Eq. (3.7), that is, at the window starting at sample number r, and with Fourier coefficients corresponding to nominal power system frequency. Using Eq. (3.7) to represent each of the phasors (adding the appropriate phase identifier as a subscript), Eq. (3.8) becomes

$$
\begin{bmatrix} X'_{r0} \\ X'_{r1} \\ X'_{r2} \end{bmatrix} = \frac{1}{3} \begin{bmatrix} 1 & 1 & 1 \\ 1 & \alpha & \alpha^2 \\ 1 & \alpha^2 & \alpha \end{bmatrix} \begin{bmatrix} PX_a \varepsilon^{jr(\omega-\omega_0)\Delta t} + QX_a{}^* \varepsilon^{-jr(\omega+\omega_0)\Delta t} \\ PX_b \varepsilon^{jr(\omega-\omega_0)\Delta t} + QX_b{}^* \varepsilon^{-jr(\omega+\omega_0)\Delta t} \\ PX_c \varepsilon^{jr(\omega-\omega_0)\Delta t} + QX_c{}^* \varepsilon^{-jr(\omega+\omega_0)\Delta t} \end{bmatrix}
\tag{3.17}
$$

or

$$
\begin{bmatrix} X'_{r0} \\ X'_{r1} \\ X'_{r2} \end{bmatrix}
$$

$$
= \frac{1}{3} \begin{bmatrix} P(X_a + X_b + X_c)\varepsilon^{jr(\omega-\omega_0)\Delta t} \\ P(X_a + \alpha X_b + \alpha^2 X_c)\varepsilon^{jr(\omega-\omega_0)\Delta t} \\ P(X_a + \alpha^2 X_b + \alpha X_c)\varepsilon^{jr(\omega-\omega_0)\Delta t} \end{bmatrix}
$$

$$
\begin{matrix} + Q(X_a{}^* + X_b{}^* + X_c{}^*)\varepsilon^{-jr(\omega+\omega_0)\Delta t} \\ + Q(X_a{}^* + \alpha X_b{}^* + \alpha^2 X_c{}^*)\varepsilon^{-jr(\omega+\omega_0)\Delta t} \\ + Q(X_a{}^* + \alpha^2 X_b{}^* + \alpha X_c{}^*)\varepsilon^{-jr(\omega+\omega_0)\Delta t} \end{matrix} \bigg].
\tag{3.18}
$$

Example 3.3   Numerical example: balanced three-phase off-nominal frequency signal.

Now consider a three-phase balanced source of frequency 60.5 Hz:

$$X_a = 100 \ \varepsilon^{j\pi/4},$$
$$X_b = 100 \ \varepsilon^{j(\pi/4 - 2\pi/3)}, \tag{3.19}$$
$$X_c = 100 \ \varepsilon^{j(\pi/4 + 2\pi/3)}.$$

This corresponds to the symmetrical components

$$X_0 = 0,$$
$$X_1 = 100 \ \varepsilon^{j\pi/4}, \tag{3.20}$$
$$X_2 = 0.$$

The first 25 samples of these signals taken at a sampling rate of 1440 Hz are given in Table 3.7.

Table 3.7 First 25 samples of the positive-sequence 60.5-Hz signal $100e^{j\pi/4}$

Sample no.	x_{ka}	x_{kb}	x_{kc}
1	100.0000	36.6025	−136.6025
2	70.4433	70.9777	−141.4210
3	36.0062	100.4354	−136.4415
4	−0.9256	122.9347	−122.0091
5	−37.7932	136.9168	−99.1235
6	−72.0424	141.4129	−69.3705
7	−101.3004	136.1117	−34.8113
8	−123.5400	121.3804	2.1597
9	−137.2205	98.2395	38.9810
10	−141.3941	68.2924	73.1017
11	−135.7716	33.6139	102.1577
12	−120.7424	−3.3935	124.1360
13	−97.3480	−40.1658	137.5139
14	−67.2091	−74.1553	141.3645
15	−32.4138	−103.0072	135.4210
16	4.6272	−124.7225	120.0953
17	41.3476	−137.7967	96.4491
18	75.2034	−141.3241	66.1207
19	103.8489	−135.0602	31.2113
20	125.2995	−119.4391	−5.8605
21	138.0691	−95.5429	−42.5262
22	141.2730	−65.0273	−76.2457
23	134.6891	−30.0065	−104.6827
24	118.7737	7.0933	−125.8670
25	94.6294	43.7016	−138.3310

A plot of the three phasor magnitudes is shown in Figure 3.11.

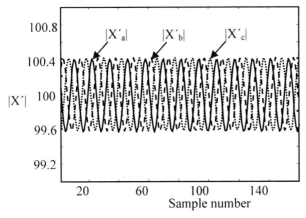

Fig. 3.11 Phasor magnitudes of individual phase quantities in balanced 60.5-Hz signals.

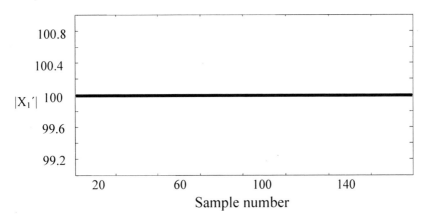

Fig. 3.12 Positive-sequence voltage magnitude estimate for a 60.5-Hz balanced signal.

It can be seen that the magnitude of errors in each of the phasors is identical to that seen in section 3.3. The magnitudes of positive-and negative-sequence components obtained from these signals are shown in Figures 3.12 and 3.13. The angle of the positive-sequence component is shown in Figure 3.14. The negative-sequence component angle is not shown in a plot since it is not very instructive; it rotates at (approximately)$2\omega_0$. The zero-sequence component is identically zero.

Figures 3.12 and 3.13 show that the positive- and negative-sequence components have no second harmonic ripple as in case of a single-phase input. The magnitudes are constant at 99.9886 and 0.4197, respectively.

This agrees with the estimates of $|P|$ and $|Q|$ (0.9999 and 0.0042 from Tables 5.1 and 5.3, respectively) at 0.5-Hz deviation from the nominal frequency.

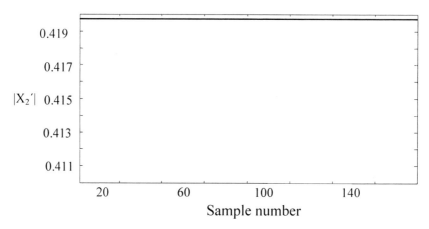

Fig. 3.13 Negative-sequence voltage magnitude estimate for a 60.5-Hz balanced signal.

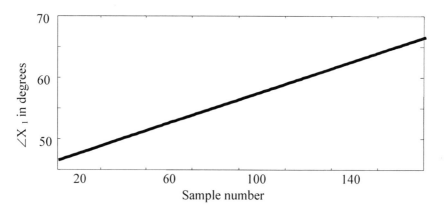

Fig. 3.14 Positive-sequence voltage angle estimate for a 60.5-Hz balanced signal.

3.5 Estimates of unbalanced input signals

Most phasor measurement applications call for measurements under normal system conditions. This usually implies balanced three-phase voltages and currents. However, it is common to have some degree of unbalance in the power system due to unbalanced loads and untransposed transmission lines. Estimates of such unbalances (negative- and zero-sequence) range between 0 and 10% of the positive-sequence component. The effect of such unbalances on positive-sequence measurement at off-nominal conditions is considered in this section.

3.5.1 Unbalanced inputs at off-nominal frequency

When the off-nominal three-phase signals are unbalanced (either or both), their negative- and zero-sequence components are non-zero. The effect of unbalances in the input signals can be best studied by considering the estimated sequence components in the presence of unbalances. The phase components in terms of their symmetrical components are

$$X_a = \{X_0 + X_1 + X_2\},$$

$$X_b = \{X_0 + \alpha^2 X_1 + \alpha X_2\}, \tag{3.21}$$

$$X_c = \{X_0 + \alpha X_1 + \alpha^2 X_2\}.$$

The symmetrical components in Eq. 3.21 are true symmetrical components of the input signals.

Substituting from Eq. (3.16) in Eq. (3.18) leads to symmetrical component estimated with nominal frequency phasor computation process:

$$
\begin{bmatrix} X'_{r0} \\ X'_{r1} \\ X'_{r2} \end{bmatrix} = \frac{1}{3}
\begin{bmatrix} P(3X_0)\varepsilon^{jr(\omega-\omega_0)\Delta t} + Q(3X^*_0)\varepsilon^{-jr(\omega+\omega_0)\Delta t} \\ P(3X_1)\varepsilon^{jr(\omega-\omega_0)\Delta t} + Q(3X^*_2)\varepsilon^{-jr(\omega+\omega_0)\Delta t} \\ P(3X_2)\varepsilon^{jr(\omega-\omega_0)\Delta t} + Q(3X^*_1)\varepsilon^{-jr(\omega+\omega_0)\Delta t} \end{bmatrix}
$$

$$
= P\varepsilon^{jr(\omega-\omega_0)\Delta t} \begin{bmatrix} X_0 \\ X_1 \\ X_2 \end{bmatrix} + Q\varepsilon^{-jr(\omega+\omega_0)\Delta t} \begin{bmatrix} X^*_0 \\ X^*_2 \\ X^*_1 \end{bmatrix}. \tag{3.22}
$$

In obtaining Eq. (3.22), use has been made of the following identities:

$$\alpha^* = \alpha^2$$
$$\alpha^{2*} = \alpha$$
$$1 + \alpha + \alpha^2 = 0 \tag{3.23}$$

Equation (3.22) displays an interesting result: at off-nominal frequencies, the positive- and negative-sequence components of the input signals create false negative- and positive-sequence components, respectively, which introduce errors in the estimate of the positive and negative components. Zero-sequence component alone makes an error contribution to the zero-sequence estimate. These error contributions, because of the multiplier Q, vanish as the frequency approaches the nominal frequency (i.e., $\omega \rightarrow \omega_0$).

Example 3.4 Numerical example: unbalanced three-phase off-nominal frequency signal.

Now consider an unbalanced input at 60.5 Hz. The three-phase inputs are assumed to be made up of the following symmetrical components:

$$\begin{aligned} X_0 &= 10\varepsilon^{j\pi/4}, \\ X_1 &= 100\varepsilon^{j\pi/4}, \\ X_2 &= 20\varepsilon^{j\pi/4}. \end{aligned} \tag{3.24}$$

The corresponding phase quantities are given by

$$\begin{aligned} X_a &= 130\varepsilon^{j0.7854}, \\ X_b &= 85.44\varepsilon^{-j1.4105}, \\ X_c &= 85.44\varepsilon^{j2.9813}. \end{aligned} \tag{3.25}$$

As before these signals are sampled at 1440 Hz and the phasor representation of their symmetrical components estimated. The first 25 samples of these input quantities are given in Table 3.8.

Table 3.8 First 25 samples of the unbalanced 60.5-Hz signals

Sample no.	x_{ka}	x_{kb}	x_{kc}
1	130.0000	19.2858	−119.2814
2	91.5763	49.7413	−120.1814
3	46.8080	76.7506	−112.7550
4	−1.2033	98.4425	−97.5167
5	−49.1312	113.3140	−75.5221
6	−93.6552	120.3349	−48.2953
7	−131.6905	119.0187	−17.7224
8	−160.6020	109.4566	14.0783
9	−178.3867	92.3111	44.9037
10	−183.8123	68.7701	72.6180

Sample no.	x_{ka}	x_{kb}	x_{kc}
11	−176.5030	40.4646	95.3012
12	−156.9652	9.3556	111.3817
13	−126.5524	−22.4016	119.7454
14	−87.3719	−52.6068	119.8129
15	−42.1380	−79.1673	111.5795
16	6.0153	−100.2428	95.6156
17	53.7519	−114.3733	73.0272
18	97.7644	−120.5798	45.3794
19	135.0036	−118.4322	14.5876
20	162.8894	−108.0794	−17.2149
21	179.4898	−90.2386	−47.8247
22	183.6548	−66.1459	−75.1211
23	175.0958	−37.4704	−97.2130
24	154.4058	−6.1989	−112.5697
25	123.0182	25.5021	−120.1273

The magnitude and angle of the symmetrical components calculated from these phasors are shown in Figures 3.15–3.20.

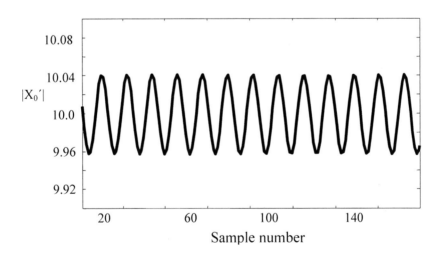

Fig. 3.15 Zero-sequence voltage magnitude estimate for a 60.5-Hz unbalanced signal.

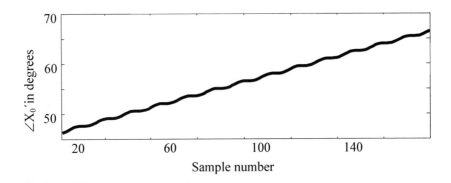

Fig. 3.16 Zero-sequence voltage angle estimate for a 60.5-Hz unbalanced signal.

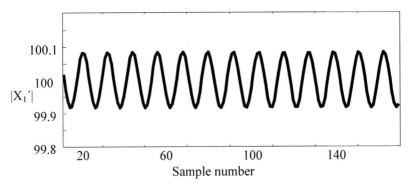

Fig. 3.17 Positive-sequence voltage magnitude estimate for a 60.5-Hz unbalanced signal.

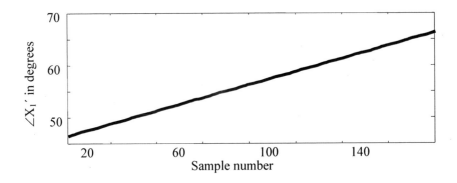

Fig. 3.18 Positive-sequence voltage angle estimate for a 60.5-Hz unbalanced signal.

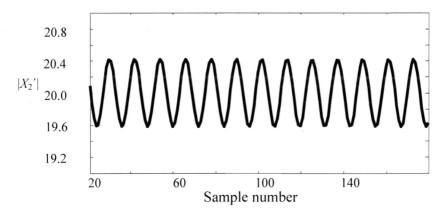

Fig. 3.19 Negative-sequence voltage magnitude estimate for a 60.5-Hz unbalanced signal.

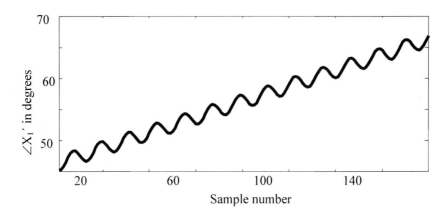

Fig. 3.20 Negative-sequence voltage angle estimate for a 60.5-Hz unbalanced signal.

Note the presence of the second harmonic ripple in all the signals. The governing equations are 3.22. The zero-sequence result (Figure 3.15) shows a steady level of 10, with a superposition of a peak-to-peak second harmonic component of 0.042, which corresponds to QX_0^*, $|Q|$ being 0.0042 for a frequency deviation of 0.5 Hz. Similarly, Figure 3.17 shows a second harmonic of 0.084 peak-to-peak in the positive-sequence estimate, which agrees with the expected value of QX_2^* with X_2 magnitude being 20. Similarly, Figure 3.19 shows the negative-sequence component estimate of a steady level of 20, with a superimposed second harmonic component of 0.42 peak-to-peak corresponding to QX_1^*, X_1 having a magnitude of 100.

All the second harmonic components can be eliminated by the three-point filter algorithm discussed in section 3.3.1. The results are similar to those shown in Figures 3.8 and 3.9.

3.5.2 A nomogram

We now summarize the result of analysis of unbalances at off-nominal frequency. The factor Q defined by Eq. (3.11) and plotted in Figure 3.5 provides the magnitude of the $(\omega + \omega_0)$ component in the output phasor when a single-phase phasor at off-nominal frequency is estimated, or when the positive-sequence phasor from an unbalanced three-phase source at off-nominal frequency is estimated. Note that in this discussion it is assumed that the off-nominal frequency phasors being measured are constant. Equations (3.9) and a part of Eq. (3.22) are reproduced here as Eqs. (3.26) and (3.27) for ready reference:

$$X_r^{'} = PX\varepsilon^{jr(\omega-\omega_0)\Delta t} + QX^*\varepsilon^{-jr(\omega+\omega_0)\Delta t}, \tag{3.26}$$

$$X_{r1}' = PX_1\varepsilon^{jr(\omega-\omega_0)\Delta t} + QX_2^*\varepsilon^{-jr(\omega+\omega_0)\Delta t}. \tag{3.27}$$

The two equations have similar forms. While the $(\omega + \omega_0)$ term in (3.26) has a magnitude proportional to QX^*, the corresponding term in Eq. (3.27) is proportional to QX_2^*. We may thus present the $(\omega + \omega_0)$ term for a single-phase phasor measurement as an error term which is a function of $\Delta\omega$, while the corresponding term for positive-sequence measurement will depend upon $\Delta\omega$ as well as upon the negative-sequence component. Normalizing the two equations,

$$X_r^{'} = X[P\varepsilon^{jr(\omega-\omega_0)\Delta t} + Q(\frac{X^*}{X})\varepsilon^{-jr(\omega+\omega_0)\Delta t}], \tag{3.28}$$

$$X_{r1}' = X_1[P\varepsilon^{jr(\omega-\omega_0)\Delta t} + Q(\frac{X_2^*}{X_1})\varepsilon^{-jr(\omega+\omega_0)\Delta t}]. \tag{3.29}$$

As noted previously, P is almost equal to 1.0 for all practical frequency excursions.

If the phase angle of the single-phase phasor X in Eq. (3.28) is θ while the phase angles of positive- and negative-sequence phasors X_1 and X_2 are θ_1 and θ_2, respectively,

$$(\frac{X^*}{X}) = \varepsilon^{-2j\theta} , \tag{3.30}$$

$$(\frac{X_2^*}{X_1}) = k_{21}\varepsilon^{-j(\theta_2+\theta_1)} , \tag{3.31}$$

where k_{21} is the ratio of magnitudes of the negative-sequence to positive-sequence component in the quantities being measured. (k_{21} represents the per unit negative-sequence component in the inputs.) Thus Eqs. (3.28) and (3.29) become

$$X_r{}' = X[P\varepsilon^{jr(\omega-\omega_0)\Delta t} +Q\varepsilon^{-jr(\omega+\omega_0)\Delta t-2j\theta}], \tag{3.32}$$

$$X_{r1}' = X_1[P\varepsilon^{jr(\omega-\omega_0)\Delta t} +Qk_{21}\varepsilon^{-jr(\omega+\omega_0)\Delta t-j(\theta_2+\theta_1)}]. \tag{3.33}$$

The magnitude of the per unit $(\omega + \omega_0)$ component in Eq. (3.32) is simply $|Q|$, whereas that in Eq. (3.33) is $|Qk_{21}|$. The single-phase measurement error component at $(\omega + \omega_0)$ can be obtained from a plot of $|Q|$ versus $\Delta\omega$ as shown in Figure 3.21.

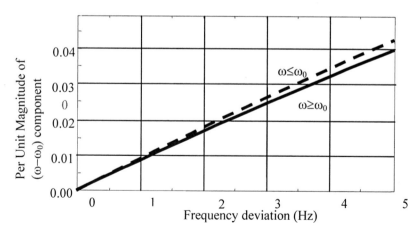

Fig. 3.21 Per unit error contribution at frequency $(\omega + \omega_0)$ when measuring a single-phase quantity with a frequency deviation of Δf.

In case of an unbalanced three-phase input, the contribution at $(\omega + \omega_0)$ is a function of two variables: Q and k_{21}. Thus the result must be presented as a curved plane in a three-dimensional plot or as a family of curves in a two-dimensional plot. These results are shown in Figures 3.22 and 3.23, respectively.

For representation of the data on a surface in three dimensions we use the expression for Q given in Eq. (3.11), and instead of $(\omega + \omega_0)$ we use the equivalent form $(2\omega_0 + \Delta\omega) = (240\pi + 2\pi\Delta f)$. Thus

$$k_{21}Q = k_{21}\left\{ \frac{\sin\dfrac{N(240\pi + 2\pi\Delta f)\Delta t}{2}}{N\sin\dfrac{(240\pi + 2\pi\Delta f)\Delta t}{2}} \right\}. \tag{3.34}$$

Further, we use the same data sampling rate as was used in earlier discussions: $N = 24$ and $\Delta t = 1/1440$ second. For this sampling rate and letting Δf vary between ± 5 Hz, leads to the surface shown in Figure 3.22.

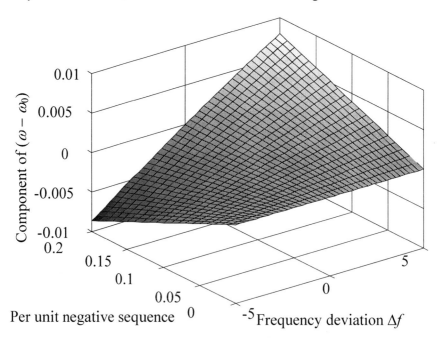

Fig. 3.22 Per unit phasor estimate error at frequency $(\omega + \omega_0)$ when measuring an unbalanced three-phase quantity with a frequency deviation of Δf. The per unit value of the negative-sequence component in terms of positive-sequence component is k_{21}. The figure is a curved surface in three dimensions representing Eq. (3.34).

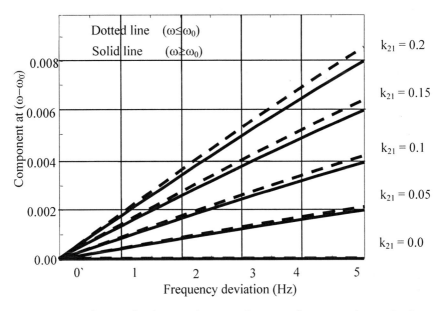

Fig. 3.23 Per unit contribution to phasor estimate at frequency $(\omega + \omega_0)$ when measuring an unbalanced three-phase quantity with a frequency deviation of Δf. The per unit value of the negative-sequence component in terms of positive-sequence component is k_{21}.

As noted previously, the factor Q changes sign with Δf. In order to simplify Figure 3.23, the dotted lines, which belong to negative Δf have been plotted with their signs reversed. The surface in Figure 3.22 does show the correct sign change in Q as Δf becomes negative.

3.6 Sampling clocks locked to the power frequency

It has been assumed thus far that the sampling process is keyed to the nominal frequency of the power system regardless of the prevailing power system frequency. As the power system is rarely at the nominal frequency, error terms appear in the phasor estimation process. The error terms were represented by the factor Q in the previous sections, and are responsible for the second harmonic ripple in the phasor magnitude and angle estimates. It has already been demonstrated above that the error terms are very small for normally expected frequency excursions in a power system, and furthermore the errors can be eliminated with the help of appropriate filtering techniques.

Another option, although not as frequently employed in phasor meas-urement unit (PMU) technology, is to track the power system frequency, and alter the sampling clock to match the period of the prevailing power system frequency [3]. A generic flow chart of such a scheme is shown in Figure 3.24. The frequency of the power system is estimated by measuring the zero crossing intervals of the voltage signals, or by using one of the frequency estimation techniques described in Chapter 4. (Alternatively, phase-locked loops may be used to track the power system frequency, and then sampling clock pulses generated at the desired sampling rate.) In any case, as the voltage waveforms may undergo step changes due to switching operations, the frequency tracking system must be designed with some care. In addition, if the voltage signal used for frequency measurement should be lost (due to a fault, or due to a blown fuse in the voltage trans-former circuit), an assured fall-back position must be available to the fre-quency tracking process.

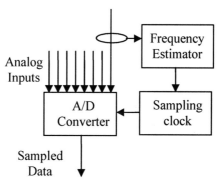

Fig. 3.24 Sampling clock synchronized to the power system frequency. The sam-pling rate (samples per cycle) is multiplied by the power system frequency to de-termine the sampling clock frequency.

If the sampling rate is matched to the power system frequency, there is no error in phasor estimation. The estimated phasor will be correct, and no ripple corresponding to the Q factor is observed. It is of course obvious that if the frequency measurement is in error, then the phasor estimate will also be in error.

An issue with this technique of phasor estimation is the correlation be-tween the measured phasor and the time-tag with which the measurement must be associated. As will be seen in the discussion of the IEEE standard which defines the requirements for PMUs, the time tags of phasor meas-urements must coincide with the Global Positioning System (GPS) second marker and with multiples of nominal periods of power system frequency measured from the GPS signal. For example, if the time-tag rate of 30 per second is selected for a 60-Hz system (i.e., once every two cycles of the nominal power system frequency), the allowable time-tags for the PMU are as shown in Figure 3.25(a). These time-tag instants are of course related to the GPS clock, and are not related to the sampling clock generated from the power system frequency measurement. If the first sample of the data window from which the phasor is estimated is as shown in Figure 3.25(b), the estimated angle of the phasor will be θ – being the angle between the first sample and the peak of the sinusoid. It then becomes necessary that the interval between the first sample of the data window and the time-tag be determined so that the correct angle corresponding to the time-tag (ϕ) be reported as the angle of the phasor.

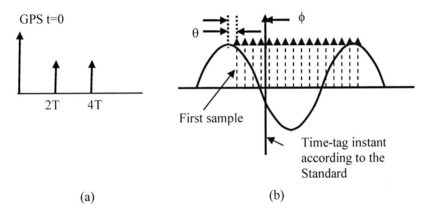

(a)   (b)

Fig. 3.25 (a) Phasor time-tags at multiples of fundamental frequency period T. Example here shows reporting rate of once every two cycles. **(b)** The phasor angle estimated by the variable frequency clock is θ, as determined by the waveform and the instant when the first sample is taken. The phase angle which must be reported in the output is ϕ.

3.7 Non-DFT-type phasor estimators

There are a number of alternative algorithms described in the literature [4–7] for estimating phasors from sampled data. Assuming that the sinusoid under consideration is given by

$$x(t) = X_1 \cos \omega t + X_2 \sin \omega t, \tag{3.35}$$

where X_1, X_2, and ω are all treated as unknown. Taking sufficient samples of the sampled data, one could now formulate the estimation problem as a nonlinear weighted least squares (WLS) problem. In addition to finding the phasors (through X_1 and X_2) the procedure would also determine the power system frequency. One could also include the DC offset in this formulation, and estimate the DC component at the same time [4].

Other techniques cited in the literature include neural networks [8], Kalman filter [9], wavelets [10] etc. The interested reader will find explanations of these methods in the cited references. In this book, we concentrate on the DFT-based techniques described in this chapter, as they provide simple, elegant, and accurate estimation of the parameters of interest.

References

1. Phadke, A.G., Thorp, J.S., and Adamiak, M.G., "A new measurement technique for tracking voltage phasors, local system frequency, and rate of change of frequency", IEEE Transactions on Power Apparatus and Systems, Vol. PAS-102, No. 5, May 1983, pp 1025–1038.
2. Phadke, A.G. and Thorp, J.S., "Improved control and protection of power systems through synchronized phasor measurements", Control and Dynamic Systems, Vol. 43, 1991 Academic Press, Inc., New York.
3. Benmouyal, G., "Design of a combined digital global differential and volt/Hertz relay for step transformer", IEEE Transactions on Power Delivery, Vol. 6, No. 3, July 1991, pp 1000–1007
4. Terzija, V.V., Djuric, M.B., and Kovacevic, B.D., "Voltage phasor and local system frequency estimation using Newton type algorithm", IEEE Transactions on Power Delivery, Volume: 9, No. 3, July 1994, pp 1368–1374.
5. Sidhu, T.S. and Sachdev, M.S, "An iterative DSP technique for tracking power system frequency and voltage phasors", Electrical and Computer Engineering, 1996. Canadian Conference on, Vol. 1, 26–29 May 1996, pp 115–118.
6. Kamwa, I. and Grondin, R, "Fast adaptive schemes for tracking voltage phasor and local frequency in power transmission and distribution systems", Transmission and Distribution Conference, 1991., Proceedings of the 1991 IEEE Power Engineering Society , 22–27 September 1991, pp 930–936
7. Jun-Zhe Y. and Chih-Wen L., "A precise calculation of power system frequency and phasor", IEEE Transactions on Power Delivery, Vol. 15, No. 2 , April 2000, pp 494–499
8. Dash, P.K., Panda, S.K., Mishra, B., and Swain, D.P. "Fast estimation of voltage and current phasors in power networks using an adaptive neural network", IEEE Transactions on Power Systems, Vol. 12, No. 4 , November 1997, pp 1494–1499
9. Girgis, A.A. and Brown, R.G., "Application of Kalman filtering in computer relaying", IEEE Transactions on Power Apparatus and Systems, Vol. PAS-100, July 1981.

10. Chi-Kong W., Ieng-Tak L., Chu-San L., Jing-Tao W., and Ying-Duo H., "A novel algorithm for phasor calculation based on wavelet analysis", Power Engineering Society Summer Meeting, 2003, IEEE, Vol. 3, 15–19 July 2001 pp. 1500–1503.

Chapter 4 Frequency Estimation

4.1 Historical overview of frequency measurement

Power system frequency measurement has been in use since the advent of alternating current generators and systems. The speed of rotation of generator rotors is directly related to the frequency of the voltages they generate. The Watt-type fly-ball governor of steam turbines (Figure 4.1) is essentially a frequency measuring device which is used in a feedback control system to keep the machine speed within a limited range around the nominal value. However, this measurement is available only at the generating stations, and there is need for measuring frequency of power system at network buses away from the generating stations.

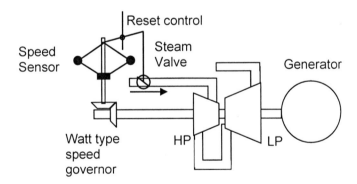

Fig. 4.1 Mechanical speed sensing used in a Watt-type speed governor of a steam turbine.

The earliest frequency measurement for power frequency voltages was performed by mechanical devices which employed mechanical resonators (similar to tuning forks) tuned to a range of frequencies around the nominal power frequency [1]. Such a frequency meter of mid-1950s vintage is shown in Figure 4.2 a. Another frequency measuring instrument of about the same period is a resonance-type device, whereby tuned resonant cir-

A.G. Phadke, J.S. Thorp, *Synchronized Phasor Measurements and Their Applications*,
DOI: 10.1007/978-0-387-76537-2_4, © Springer Science+Business Media, LLC 2008

cuits at different frequencies are energized by the secondary voltage obtained from a voltage transformer, and the circuit which is in resonance provides the frequency measurement (Figure 4.2b) [1]. Typical resolution of these meters was of the order of 0.25 Hz.

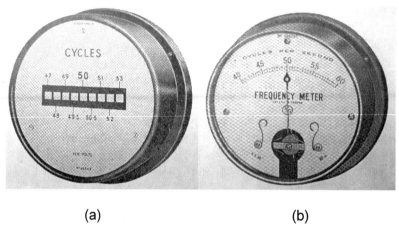

(a)   (b)

Fig. 4.2 (a) A mechanical resonance-type frequency meter. **(b)** An electrical resonance-type frequency meter. These instruments are for a 50-Hz power system.

The next advance in frequency measurement came with the introduction of precise time measurement techniques. By measuring the time interval between consecutive zero crossings of the voltage waveform the frequency of the voltage could be determined. Clearly the accuracy of such a measurement depends upon the precision of time measurement, as well as on the accuracy with which the zero crossing of the waveform could be determined. This latter measurement is affected by the presence of noise in the measurement, varying harmonic frequencies and levels, and the performance of the zero-crossing detector circuits.

Synchronized phasor measurements offer an opportunity for measuring power system frequency which eliminates many of these error sources. It should be noted that the frequency measurement on a power system is primarily dedicated to estimating rotor speed(s) of connected generators. As such, the positive-sequence voltage measurement is an ideal vehicle for frequency measurement. In addition, phasors reflect the fundamental frequency components of the voltages, and harmonics do not affect frequency measurement based upon phasors. Techniques for measuring frequency from phasors are described in the following sections.

4.2 Frequency estimates from balanced three-phase inputs

Frequency and rate of change of frequency can be estimated from the phase angles of phasor estimates [2]. It was pointed out in Chapter 3 that positive-sequence phasor estimated from balanced inputs at off-nominal frequencies has a minor attenuation in phasor magnitude, and both the magnitude and phase angle estimates are free from a ripple of approximate second harmonic. Setting the negative-sequence component of the input $X_2 = 0$ in Eq. (3.22) the estimate of the positive-sequence voltage is given by

$$X'_{r1} = PX_1 \varepsilon^{jr(\omega - \omega_0)\Delta t} . \tag{4.1}$$

The magnitude of P is the attenuation factor, and phase angle of P is a constant offset in the measured phase angles. The angle of the phasor X'_{r1} advances at each sample time by $(\omega - \omega_0)\Delta t$, where ω is the signal frequency, ω_0 is the nominal system frequency, and Δt is the sampling interval.

It should be clear from Eq. (4.1) that the first and second derivatives of phase angle of the phasor estimate would provide an estimate of $\Delta\omega = (\omega - \omega_0)$, and the rate of change of frequency. Since there are errors of estimation in phasor calculation, it is desirable to use a weighted least-squares approach over a reasonable data window for calculating the derivatives of the phase angle.

Assume that the positive-sequence phasors are estimated over one period of the nominal frequency, and that the phasors calculated with several consecutive data windows over a span of 3–6 cycles are used for frequency and rate of change of frequency estimation.

Let $[\phi_k]$ $\{k = 0,1,\dots,N - 1\}$ be the vector of "N" samples of the phase angles of the positive-sequence measurement. The vector $[\phi_k]$ is assumed to be monotonically changing over the window of "N" samples. As the phase angles of the phasor estimate may be restricted to a range of 0–2π, it may be necessary to adjust the angles to make them monotonic over the entire spanning period by correcting any offsets of 2π radians which may exist. This is illustrated in Figure 4.3.

If the frequency deviation from the nominal value, and the rate of change of frequency at $t = 0$ are $\Delta\omega$ and ω', respectively, the frequency at any time "t" is given by

$$\omega(t) = (\omega_0 + \Delta\omega + t\omega'). \tag{4.2}$$

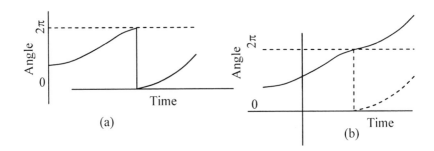

Fig. 4.3 (a) Phasor estimates produce angles which are restricted to a range of 0–2π. **(b)** For estimating frequency and rate of change of frequency, the offset of 2π in the phase angle estimates are removed.

The phase angle is the integral of the frequency:

$$\phi(t) = \int \omega dt = \int (\omega_0 + t\Delta\omega + t\omega')dt = \phi_0 + t\omega_0 + t\Delta\omega + \frac{1}{2}t^2\omega', \quad (4.3)$$

ϕ_0 being the initial value of the angle. Assuming that the recursive algorithm is used for estimating the phasors, the term $t\omega_0$ is suppressed from the estimated phase angles (see Section 2.2.2). Thus, the phase angle as a function of time becomes

$$\phi(t) = \phi_0 + t\Delta\omega + \frac{1}{2}t^2\omega'. \quad (4.4)$$

If $\phi(t)$ is assumed to be a second degree polynomial of time:

$$\phi(t) = a_0 + a_1 t + a_2 t^2 \quad (4.5)$$

it follows that at $t = 0$,

$$\Delta\omega = a_1,$$
$$\omega' = 2a_2, \quad (4.6)$$

or, in terms of Hz and Hz s^{-1},

$$\Delta f_0 = a_1/(2\pi) \text{ and } f' = a_2/(\pi). \quad (4.7)$$

The vector of "N" angle measurements is given by

$$\begin{bmatrix} \phi_0 \\ \phi_1 \\ \phi_2 \\ \vdots \\ \phi_{N-1} \end{bmatrix} = \begin{bmatrix} 1 & 0 & 0 \\ 1 & \Delta t & \Delta t^2 \\ 1 & 2\Delta t & 2^2\Delta t^2 \\ \vdots & \vdots & \vdots \\ 1 & (N-1)\Delta t & (N-1)^2\Delta t^2 \end{bmatrix} \begin{bmatrix} a_0 \\ a_1 \\ a_2 \end{bmatrix}. \tag{4.8}$$

In matrix notation

$$[\phi]=[B][A], \tag{4.9}$$

where $[B]$ is the coefficient matrix in Eq. (4.8). The unknown vector $[A]$ is calculated by the weighted least-squares (WLS) technique:

$$[A] = [B^T B]^{-1} B^T [\phi] = [G] [\phi], \tag{4.10}$$

where

$$[G] = [B^T B]^{-1} B^T \tag{4.11}$$

The matrix $[G]$ is pre-calculated and stored for use in real time. It has "N" rows and three columns. In real time, $[G]$ is multiplied by $[\phi]$ to obtain the vector $[A]$, and from that the frequency and rate of change of frequency at any time t (which is a multiple of Δt) can be calculated. This time is usually associated with the time tag for which the measurement is posted.

Example 4.1
Numerical example of frequency and rate of change of frequency estimation.

Consider an input with a frequency of 60.5 Hz and a rate of change of frequency of 1 Hz s^{-1}. The polynomial for phase angles is given by

$$\phi(t) = \phi_0 + 2\pi\times0.5\times t + (1/2)t^2\times1\times2\pi.$$

The initial angle ϕ_0 is assumed to be 0.1 radian. Assuming that the phasors are calculated at a sampling rate of 24 samples per cycle of the nominal power system frequency, the time step is $\Delta t = (1/1440)$ second. Phase angles over a span of four cycles are tabulated below with and without a Gaussian random noise with zero mean and a standard deviation of 0.01 radian (Table 4.1).

Table 4.1 Partial list of 96 phase angle samples with and without noise

Sample no.	Phase angles without noise	Phase angles with noise
1	0.1000	0.1007
2	0.1022	0.1017
3	0.1044	0.1062
4	0.1066	0.1077
5	0.1088	0.1075
6	0.1109	0.1103
...	...	...
88	0.3013	0.3015
89	0.3037	0.3033
90	0.3062	0.3056
91	0.3086	0.3087
92	0.3111	0.3123
93	0.3135	0.3110
94	0.3160	0.3166
95	0.3185	0.3175
96	0.3209	0.3219

The estimated frequency and rate of change of frequency using the weighted least-squares formulation is found to be $\Delta f = 0.5000$ Hz and $f'(0) = 1.0000$ Hz s^{-1} with no noise in the phase angle measurement, and $\Delta f = 0.4968$ Hz and $f'(0) = 1.0550$ Hz s^{-1} with the noise.

The estimates of Δf and f' for different amounts of random noise in phase angle measurements are shown in Table 4.2.

Table 4.2 Effect of random noise on frequency and rate of change of frequency estimation

σ of random noise (radians)	Mean of frequency estimate (Hz)	σ of frequency estimate(Hz)	Mean of rate of change of frequency estimate(Hz s^{-1})	σ of rate of change of frequency estimate (Hz)
0.0001	0.5000	0.0003	1.0000	0.0096
0.0005	0.5001	0.0017	0.9979	0.0498
0.0010	0.5000	0.0033	0.9986	0.0982
0.0050	0.5000	0.0167	1.0038	0.4888
0.0100	0.4998	0.0331	1.0033	0.9660

The results in Table 4.2 are for 1000 Monte Carlo trials with the specified standard deviation of the noise. It is clear that the rate of change of frequency is more sensitive to the amount of noise in the input. Also, the estimates are essentially zero-mean processes.

4.3 Frequency estimates from unbalanced inputs

The effect of unbalance in input signals has been analyzed in Section 3.5. Equation (3.22) provides the formula for the estimate of the positive-sequence component when there is a negative-sequence component present in the input signal:

$$X'_{r1} = PX_1 \varepsilon^{jr(\omega-\omega_0)\Delta t} + QX_2^* \varepsilon^{-jr(\omega+\omega_0)\Delta t}, \tag{4.12}$$

where Q is given by Eq. (3.11), and X_2 is the negative-sequence component in the input signals. The effect of the second term of Eq. (4.12) is to produce a ripple in the angle estimate of the positive-sequence component. This ripple can be eliminated by one of the filtering techniques described in Section 3.3. When the ripple in angle is eliminated, the frequency and rate of change of frequency can be estimated as in Section 4.2. The error performance of the estimates is then identical to that corresponding to balanced input signals.

4.4 Nonlinear frequency estimators

It is possible to formulate the frequency and rate of change of frequency estimation problem as a nonlinear estimation problem from the input signal waveform [3, 4]. Consider a single-phase input having a frequency deviation of $\Delta\omega$ and a rate of change of frequency ω' (as in Eq. 4.4):

$$x(t) = X \cos\left\{\phi_0 + t\Delta\omega + \frac{1}{2}t^2\omega'\right\}. \tag{4.13}$$

"N" samples of this signal at a sampling interval of Δt are $\{x_k, k = 0,1,...,N - 1\}$. It is assumed that there are four unknowns in the data samples:

$$z = \begin{bmatrix} X \\ \phi_0 \\ \Delta\omega \\ \omega' \end{bmatrix}. \tag{4.14}$$

The function $x(t)$ is a nonlinear function of the four unknowns, and if "N" is greater than four, a nonlinear weighted least-squares iterative technique can be used to solve for the four unknowns.

Assuming reasonable initial values of the four unknowns as $[z_0]$, the initial estimates of the function $x(t)$ are $[x_0]$. Using first-order terms of Taylor's series to represent the nonlinear function around $[z_0]$

$$[x - x_0] = \left[\frac{\partial x}{\partial X} \quad \frac{\partial x}{\partial \phi_0} \quad \frac{\partial x}{\partial \Delta \omega} \quad \frac{\partial x}{\partial \omega'} \right]_{z=z_0} [\Delta z]., \qquad (4.15)$$

where the partial derivatives are columns of "N" rows evaluated at the assumed value of the unknown vector $[z_0]$. Representing the matrix of partial derivatives by the Jacobian matrix $[J]$, the weighted least-squares solution for $[\Delta z]$ is

$$[\Delta z] = [J^T J]^{-1} J^T [x - x_0]. \qquad (4.16)$$

The four partial derivatives in Eq. (4.15) are obtained by differentiating the expression for $x(t)$:

$$J_1 = \frac{\partial x}{\partial X} = \cos(\phi_0 + t\Delta\omega + \frac{1}{2}t^2\omega'),$$

$$J_2 = \frac{\partial x}{\partial \phi_0} = -X \sin(\phi_0 + t\Delta\omega + \frac{1}{2}t^2\omega'),$$

$$J_3 = \frac{\partial x}{\partial \Delta\omega} = -Xt \sin(\phi_0 + t\Delta\omega + \frac{1}{2}t^2\omega'), \qquad (4.17)$$

$$J_4 = \frac{\partial x}{\partial \omega'} = -X \frac{t^2}{2} \sin(\phi_0 + t\Delta\omega + \frac{1}{2}t^2\omega').$$

Having calculated the corrections $[\Delta z]$ in Eq. (4.16), they are added to $[z_0]$ to produce the answer at the end of first iteration. The process is repeated until the residual $[x - x_0]$ becomes smaller than a suitable tolerance.

Example 4.2 Numerical example of nonlinear frequency and rate of change of frequency estimation.

Consider a single-phase input with an amplitude of 1.1, a frequency 60.5 Hz at $t = 0$, a rate of change of frequency of 1 Hz s^{-1}, and a phase angle ϕ_0 of $\pi/8$. The initial values for starting the iteration are assumed to be

$$X = 1.0,$$
$$\phi_0 = 0,$$
$$\Delta\omega = 0,$$
$$\omega' = 0.$$

The sampling rate is assumed to be 1440 Hz, and 96 samples of the input signal are used to estimate the signal parameters.

Table 4.3 lists first 10 values of the input signal, the estimated signal with the vector $[z_0]$, and the first 10 entries of the Jacobian matrix.

Table 4.3 First 10 values of $[x]$, $[x_0]$, and $[J]$ at the beginning of the iteration

	$[x]$	$[x_0]$	$[J_1]$	$[J_2]$	$[J_3]$	$[J_4] \times 10^4$
1	1.0163	1.0000	1.0000	0	0	0
2	0.8712	0.9659	0.9659	−0.2588	−0.0002	−0.0006
3	0.6658	0.8660	0.8660	−0.5000	−0.0007	−0.0048
4	0.4143	0.7071	0.7071	−0.7071	−0.0015	−0.0153
5	0.1340	0.5000	0.5000	−0.8660	−0.0024	−0.0334
6	−0.1555	0.2588	0.2588	−0.9659	−0.0034	−0.0582
7	−0.4343	0.0000	0.0000	−1.0000	−0.0042	−0.0868
8	−0.6830	−0.2588	−0.2588	−0.9659	−0.0047	−0.1141
9	−0.8843	−0.5000	−0.5000	−0.8660	−0.0048	−0.1336
10	−1.0244	−0.7071	−0.7071	−0.7071	−0.0044	−0.1381

The corrections vector at the end of first iteration are

$$\begin{aligned}
\Delta(X) &= -0.0365, \\
\Delta(\phi_0) &= 0.4072, \\
\Delta(\Delta\omega) &= 4.2767, \\
\Delta(\omega') &= -32.2073.
\end{aligned}$$

At the end of four iterations correct values for the unknowns are obtained.

It is necessary to consider the effect of noise in the sampled data on the performance of the nonlinear frequency estimator. Zero-mean normally distributed random noise was added to the sampled data of the above example, and the effect on the results obtained after five iterations evaluated. Thousand Monte Carlo trials produce the result shown in Table 4.4. It can be seen that the mean of the parameter estimation is very close to the true value, although the standard deviation of the estimation increases very rapidly as the size of the noise added increases. The rate of change of frequency is practically unusable when the noise exceeds 1% of the signal peak value. The amplitude and phase angle estimates are quite good even for very large sample errors.

Table 4.4 Effect of sample noise on estimation of signal parameters

σ sample noise	Mean X	Mean ϕ_0	Mean $\Delta\omega$	Mean ω'	σX	$\sigma\phi_0$	$\sigma\Delta\omega$	$\sigma\omega'$
0.0	1.1000	0.3927	0.5000	1.0000	0	0	0	0
0.01	1.1000	0.3929	0.4982	1.0511	0.0014	0.0037	0.0430	1.3061
0.02	1.1000	0.3924	0.5011	0.9927	0.0028	0.0076	0.0878	2.6582
0.03	1.1001	0.3923	0.5028	0.8986	0.0043	0.0112	0.1323	4.0170
0.04	1.1001	0.3926	0.4992	1.0135	0.0056	0.0144	0.1620	4.9061
0.05	1.1001	0.3928	0.5045	0.8704	0.0072	0.0187	0.2153	6.4638
0.06	1.1007	0.3930	0.5068	0.7559	0.0086	0.0223	0.2543	7.7155
0.07	1.1004	0.3930	0.5028	0.8911	0.0099	0.0269	0.3029	9.0593
0.08	1.1010	0.3910	0.5185	0.4604	0.0119	0.0293	0.3424	10.4239
0.09	1.1003	0.3942	0.4814	1.6546	0.0129	0.0337	0.3877	11.7737
0.10	1.1006	0.3943	0.4957	1.0495	0.0143	0.0376	0.4451	13.5048

The signals are the same as for Example 4.1.

4.5 Other Techniques for frequency measurements

A number of other techniques for measuring power system frequency have been published in the technical literature [5, 6, 7, 8]. These references are provided for the interested reader as a sample of what is available, and is by no means a complete listing of papers dealing with frequency measurement. In general, the faster approaches (measurements made within one or two periods of the power frequency signal) tend to have greater errors than those using longer data windows. It is better to keep in mind that a traditional use of frequency measurement is in under-frequency load-shedding. Relays used for that purpose tend to have operating times of the order of 5–6 cycles of the nominal power frequency. This is probably a good size for a data window to be used in frequency estimation.

One should not have excessively long data windows for frequency estimation in order to improve the accuracy of the estimate. During transient stability swings, the frequency of the power system may change rapidly. Thus, a long window may include significantly different frequencies over the window span, and once again the frequency estimation may be in error. We will consider the effect of changing frequency due to transients in Chapter 6.

References

1. Lythall, R.T, "The J. & P. Switchgear Book", Johnson & Phillips Ltd., Charlton, London, S.E. 7. 5th Edition, 1953., pp 441–442.
2. Phadke, A.G., Thorp, J.S., and Adamiak, M.G., "A new measurement technique for tracking voltage phasors, local system frequency, and rate of change of frequency", IEEE Transactions on Power Apparatus and Systems, Vol. 102, No. 5, pp 1025–1038.
3. Sachdev, M.S. and Giray, M.M., "A least error squares technique for determining power system frequency", IEEE Transactions on Power Apparatus and Systems Volume PAS-104, No. 2, February 1985, pp 437–444.
4. Terzija, V.V., Djuric, M.B., and Kovacevic, B.D., "Voltage phasor and local system frequency estimation using Newton-type algorithms", IEEE Transactions on Power Delivery, Vol. 9, No. 3, 1994, pp 1368–1374.
5. Sidhu, T.S. and Sachdev, M.S., "An iterative technique for fast and accurate measurement of power system frequency", IEEE Transactions on Power Delivery, Vol. 13, No.1, 1998, pp 109–115.
6. Girgis, A.A. and Hwang, T.L.D., "Optimal estimation of voltage phasors and frequency deviation using linear and non-linear Kalman filtering", IEEE Transactions on Power Apparatus and Systems, Vol. 103, No. 10, 1984, pp 2943–2949.
7. Moore, P.J., Carranza, R.D., and Johns, A.T., "A new numeric technique for high-speed evaluation of power system frequency", IEEE Proceedings-Generation, Transmission and Distribution, Vol. 141, No. 5, 1994, pp 529–536.
8. Hart, D., Novosel, D., Hu, Y., Smith, B., and Egolf,M., "A new frequency tracking and phasor estimation algorithm for generator protection", Paper No. 96, SM 376-4-PWRD, 1996, IEEE-PES Summer Meeting, Denver, July 28 – August 1, 1996

Chapter 5 Phasor Measurement Units and Phasor Data Concentrators

5.1 Introduction

The history of phasor measurement unit (PMU) evolution was discussed in Chapter 1. In this chapter we will consider certain practical implementation aspects of the PMUs and the architecture of the data collection and management system necessary for efficient utilization of the data provided by the PMUs. One of the most important features of the PMU technology is that the measurements are time-stamped with high precision at the source, so that the data transmission speed is no longer a critical parameter in making use of this data. All PMU measurements with the same time-stamp are used to infer the state of the power system at the instant defined by the time-stamp. It is clear that PMU data could arrive at a central location at different times depending upon the propagation delays of the communication channel in use. The time-tags associated with the phasor data provide an indexing tool which helps create a coherent picture of the power system out of such data. The Global Positioning System (GPS) has become the method of choice for providing the time-tags to the PMU measurements, and will be described briefly in the following sections. Other aspects of the overall PMU data collection system such as "phasor data concentrators" (PDCs), communication systems, etc. will also be considered in this Chapter.

The industry standards which define file structures for compliant PMUs have been very important to ensure interoperability of PMUs made by different manufacturers, and will be considered in section 5.6.

5.2 A generic PMU

The PMUs manufactured by different manufacturers differ from each other in many important aspects. It is therefore difficult to discuss the PMU hardware configuration in a way which is universally applicable. However,

A.G. Phadke, J.S. Thorp, *Synchronized Phasor Measurements and Their Applications*,
DOI: 10.1007/978-0-387-76537-2_5, © Springer Science+Business Media, LLC 2008

it is possible to discuss a generic PMU, which will capture the essence of its principal components.

Figure 5.1 is based upon the configuration of the first PMUs built at Virginia Tech (and shown in Figure 1.1). Remember that PMUs evolved out of the development of the "symmetrical component distance relay". Consequently the structure shown in Figure 5.1 parallels that of a computer relay. The analog inputs are currents and voltages obtained from the secondary windings of the current and voltage transformers. All three-phase currents and voltages are used so that positive-sequence measurement can be carried out. In contrast to a relay, a PMU may have currents in several feeders originating in the substation and voltages belonging to various buses in the substation.

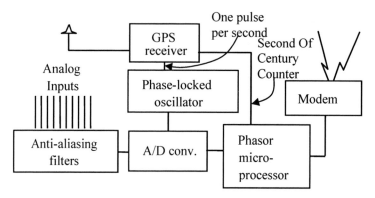

Fig. 5.1 Major elements of the modern PMU. All elements of the PMU with the exception of the GPS receiver are to be found in computer relays as well.

The current and voltage signals are converted to voltages with appropriate shunts or instrument transformers (typically within the range of ±10 volts) so that they are matched with the requirements of the analog-to-digital converters. The sampling rate chosen for the sampling process dictates the frequency response of the anti-aliasing filters. In most cases these are analog-type filters with a cut-off frequency less than half the sampling frequency in order to satisfy the Nyquist criterion. As in many relay designs [1] one may use a high sampling rate (called oversampling) with corresponding high cut-off frequency of the analog anti-aliasing filters. This step is then followed by a digital 'decimation filter' which converts the sampled data to a lower sampling rate, thus providing a 'digital anti-aliasing filter' concatenated with the analog anti-aliasing filters. The advantage of such a scheme is that the effective anti-aliasing filters made up of an analog front end and a digital decimation filter are far more stable as far as aging and temperature variations are concerned. This ensures that all the analog signals have the same phase shift and attenuation, thus assuring that the phase angle differences and relative magnitudes of the different signals are unchanged.

As an added benefit of the oversampling technique, if there is a possibility of storing raw data from samples of the analog signals, they can be of great utility as high-bandwidth "digital fault recorders".

The sampling clock is phase-locked with the GPS clock pulse (to be described in the following section). Sampling rates have been going up steadily over the years – starting with a rate of 12 samples per cycle of the nominal power frequency in the first PMUs to as high as 96 or 128 samples per cycle in more modern devices, as faster analog-to-digital converters and microprocessors have become commonplace. Even higher sampling rates are certainly likely in the future leading to more accurate phasor estimates, since higher sampling rates do lead to improved estimation accuracy [1].

The microprocessor calculates positive-sequence estimates of all the current and voltage signals using the techniques described in Chapters 2–4 earlier. Certain other estimates of interest are frequency and rate of change of frequency measured locally, and these also are included in the output of the PMU. The time-stamp is created from two of the signals derived from the GPS receiver. This will be considered in greater detail in the next section. For the moment, it is sufficient to say that the time-stamp identifies the identity of the "universal time coordinated (UTC) second and the instant defining the boundary of one of the power frequency periods as defined in the IEEE standard to be considered in section 5.6 below.

Finally, the principal output of the PMU is the time-stamped measurement to be transferred over the communication links through suitable modems to a higher level in the measurement system hierarchy. It is the specification of these output file structures which is the subject of the industry standard for PMUs to be considered in section 5.6.

5.3 The global positioning system

The GPS was initiated with the launch of the first Block I satellites in 1978 by US Department of Defense.[1] By 1994 the complete constellation of 24 modern satellites was put in place. (In 2007 there are 30 active satellites in

---

[1] There is great wealth of information about the GPS system available in various technical publications. A highly readable account for the layman is available at the web-site http://wikipedia.com. There the interested reader will also find links to other source material.

orbit, the extra satellites providing for greater accuracy in estimation of spatial coordinates of the receivers. Block I and II satellites have been retired.) These are arranged in six orbital planes displaced from each other by 60° and having an inclination of about 55° with respect to the equatorial plane (see Figure 5.2). The satellites have an orbital radius of 16,500 miles, and go around the earth twice during one day. They are so arranged that at least six satellites are visible at most locations on earth, and often as many as 10 satellites may be available for viewing. The most common use of the GPS system is in determining the coordinates of the receiver, although for the PMUs the signal which is most important is the one pulse-per-second. This pulse as received by any receiver on earth is coincident with all other received pulses to within 1 microsecond. In practice much better accuracies of synchronization – of the order of a few hundred nano-seconds – have been realized.

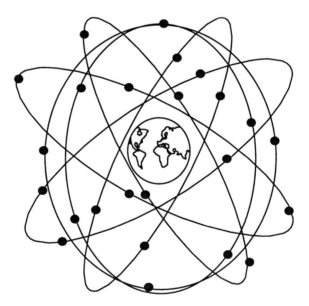

Fig. 5.2 Representation of the GPS satellite disposition. There are four satellites in each of the six orbits, which orbit around the earth with a period of half a day.

The GPS satellites keep accurate clocks which provide the one pulse-per-second signal. The time they keep is known as the GPS time which does not take into account the earth's rotation. Corrections to the GPS time are made in the GPS receivers to account for this difference (leap-second correction) so that the receivers provide UTC clock time. The identity of the pulse is defined by the number of seconds since the time that the clocks began to count (January 6, 1980). It should be noted that the PMU standard (see section 5.6) uses UNIX time base with a "second-of-century" (SOC) counter which began counting at midnight on January 1, 1970.

At present there are a number of GPS-like systems being deployed by other nations, with similar goals. It is expected that the GPS system will remain the principal source of synchronization for PMUs for the foreseeable future.

5.4 Hierarchy for phasor measurement systems

The PMUs are installed in power system substations. The selection of substations where these installations take place depends upon the use to be made of the measurements they provide. The optimal placement of PMUs will be considered in some of the following chapters which discuss some of the applications of phasor measurements.

In most applications, the phasor data is used at locations remote from the PMUs. Thus an architecture involving PMUs, communication links, and data concentrators must exist in order to realize the full benefit of the PMU measurement system. A generally accepted architecture of such a system is shown in Figure 5.3.

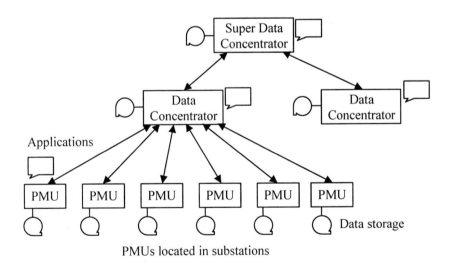

Fig. 5.3 Hierarchy of the phasor measurement systems, and levels of phasor data concentrators.

In Figure 5.3 the PMUs are situated in power system substations, and provide measurements of time-stamped positive-sequence voltages and currents of all monitored buses and feeders (as well as frequency and rate of change of frequency). The measurements are stored in local data storage devices, which can be accessed from remote locations for post-mortem or diagnostic purposes. The local storage capacity is necessarily limited, and the stored data belonging to an interesting power system event must be flagged for permanent storage so that it is not overwritten when the local storage capacity is exhausted. The phasor data is also available for real-time applications in a steady stream as soon as the measurements are made. There may well be some local application tasks which require PMU data, in which case it can be made available locally for such tasks. However, the main use of the real-time data is at a higher level where data from several PMUs is available.

The devices at next level of the hierarchy are commonly known as "phasor data concentrators" (PDCs). Typical function of the PDCs is to gather data from several PMUs, reject bad data, align the time–stamps, and create a coherent record of simultaneously recorded data from a wider part of the power system. There are local storage facilities in the PDCs, as well as application functions which need the PMU data available at the PDC. This can be made available by the PDCs to the local applications in real time. (Clearly, the communication and data management delays at the PDCs will create greater latency in the real-time data, but all practical experience shows that this is not unmanageable. The question of data latency will be further considered when applications are discussed in later chapters.)

One may view the first hierarchical level of PDCs as being regional in their data-gathering capability. On a systemwide scale, one must contemplate another level of the hierarchy (Super Data Concentrator in Figure 5.3). The functions at this level are similar to those at the PDC levels – that is, there is facility for data storage of data aligned with time-tags (at a somewhat increased data latency), as well as a steady stream of near real-time data for applications which require data over the entire system.

Figure 5.3 shows the communication links to be bidirectional. Indeed, most of the data flow is upward in the hierarchy, although there are some tasks which require communication capability in the reverse direction. Usually these are commands for configuring the downstream components, requesting data in a particular form, etc. In general, the capacity for downward communication is not as demanding as one in the upward direction. These issues will be considered in Section 5.6 where the prevailing industry standard for PMUs [2] will be discussed.

5.5 Communication options for PMUs

Communication facilities are essential for applications requiring phasor data at remote locations. Two aspects of data transfer are significant in any communication task [3, 4]. Channel capacity is the measure of the data rate (in kilobits per second or megabits per second) that can be sustained on the available data link. The second aspect is the latency, defined as the time lag between the time at which the data is created and when it is available for the desired application. The data volume created by the PMUs is quite modest, so that channel capacity is rarely a limiting factor in most applications. On the other hand, some applications may require relatively small latency – in particular, applications for real-time control of power systems. At the other extreme are post-mortem analysis applications, which require PMU data to help analyze the power system performance during major disturbances. These applications are not affected by large delays in transferring the data. Several applications of PMU data will be considered in the following chapters.

The communication options available for PMU data transfer may be classified according to the physical medium used for communication [5]. Leased telephone circuits were among the first communication media used for these purposes. Switched telephone circuits can be used when data transfer latency is not of importance. More common electric utility communication media such as "power line carrier" and microwave links have also been used, and continue to be used in many current applications. Of course, the medium of choice now is fiber-optic links which have unsurpassed channel capacity, high data transfer rates, and immunity to electromagnetic interference. Figure 5.4 [5] shows typical construction of a fiber-optic cable

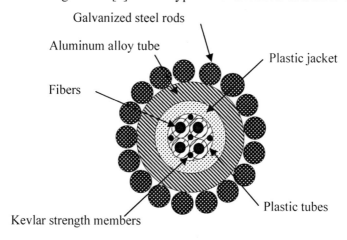

Fig. 5.4 Typical fiber optic cable construction. Such cables are in wide use on electric utility systems.

commonly used in electric utility industry. The most popular deployment of fiber-optic cables is in the ground wires of transmission lines. The ground wires may carry multiple fibers which may be used for other communication, protection, and control applications for power system operation and management. Other configurations of fiber-optic links may involve separate towers for the fiber cable in the electric utility right-of-way, wrapping the fiber cable around the phase conductors, or direct burying the fiber cable in the ground (see Figure 5.5).

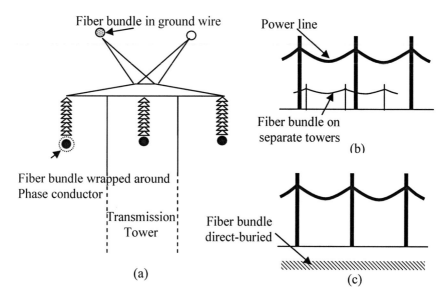

Fig. 5.5 Arrangement of fiber-optic bundle commonly employed by electric utilities. (**a**) The fiber is in the ground wire. (**b**) The fiber bundle is strung on separate towers on transmission line right-of-way. (**c**) The fiber-optic cable is direct-buried.

The technology of fiber-optics is changing very rapidly [6]. Fibers may be single-mode (meaning that the entire fiber cross-section is homogeneous material) or multimode with graded index or step-index change in the refractive index of the fiber and the cladding material (see Figure 5.6). Multimode fibers tend to have greater loss per km, because of the partial loss of energy due to refraction at the boundary between the fiber core and the cladding. Single-mode fibers propagate optical waves along the axis of the fiber, and have minimal loss during transmission. The wavelength of light used in these systems ranges from 900 nm to 1800 nm. Typical loss in the fibers may range from 0.5 db per km (single-mode fibers) to 4 db per km (multimode fibers). In addition, loss in the connectors and repeaters

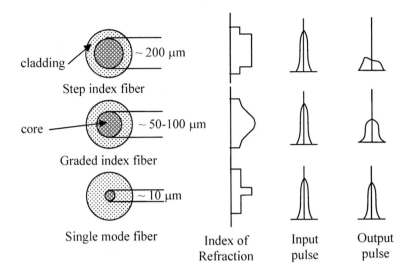

cladding   ~ 200 μm

Step index fiber

core   ~ 50-100 μm

Graded index fiber

~ 10 μm

Single mode fiber

Index of
Refraction

Input
pulse

Output
pulse

Fig. 5.6 Types of fibers used in fiber-optic communication, their relative dimensions, and modes of transmission.

must also be taken into account. Depending upon the length of transmission path and allowed transmission loss-budget, an appropriate type of fiber is selected for a given application.

One may also classify the communication facilities based upon the communication protocols in use. Here also the field is rapidly changing, and it is only possible to mention a few of the available protocols which have been used in phasor measurement applications. The IEEE standard applicable to PMU technology [2] discusses the general requirements for communications with PMUs in Annexe I of the standard. When serial communication over an RS-232 is used the entire data stream from the PMU (as defined in the PMU Standard and discussed more fully in the next section) is to be mapped in proper order on to the serial communication port. The communication system may apply any protocol, encryption, or change the ordering of the data, as long as it is restored to its original format at the receiver end.

PMU messages may also be mapped in their entity into transmission control protocol (TCP) [7] or user datagram protocol (UDP) [8], and will be accessed by using standard Internet Protocol (IP) functions. The (IP) may be carried over Ethernet or other available transport means.

In recent years IEC Standard 61850 has been introduced to facilitate electric utility substation automation including protection and control [9]. In its present version, this standard has not been identified as being useable for PMUs. It may be that advances in the state-of-the-art in PMU technology and substation communication technology will lead to the acceptance of IEC 61850 by the PMU community.

5.6 Functional requirements of PMUs and PDCs

5.6.1 The evolution of "Synchrophasor" standard

In order to achieve interoperability among PMUs made by different manufacturers, it is essential that all PMUs perform to a common standard. Reference [7] is the current IEEE standard which defines requirements for compliance.

A short account of the development of this standard may be of interest. The "Synchrophasor" standard was first issued in 1995 [10]. PMUs of early manufacture based on this standard were tested for interoperability, and it was discovered that their performance at off-nominal frequencies was not identical [11]. From the point of interoperability of equipment, this was not acceptable. It was soon recognized that the then existing standard [11] was not very clear on the topic of performance requirements for PMUs at off-nominal frequencies. A working group of the Power System Relaying Committee of IEEE undertook the revision of the standard, and the result is the current standard [2] which clarified the requirements for PMU response to off-nominal frequency inputs. The requirements for off-nominal frequencies can be explained with the help of Figure 5.7. As noted in Chapter 1 the definition of a phasor is independent of its frequency; thus, if the input signals connected to the PMU are pure sinusoids of any frequency and the phasor estimate is reported at the time-tag as shown in Figure 5.7, the output phasor must have a magnitude equal to the rms value

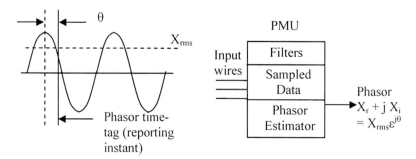

Fig. 5.7 PMU performance requirements for input signals of any frequency. (**a**) Input signal connected to the PMU terminals, and (**b**) the required output phasor estimate.

of the signal, and its phase angle must be θ, the angle between the reporting instant and the peak of the sinusoid. Note that the PMUs in general contain a number of filters at the input stage. The phase delays caused by these filters must be compensated for before the phasor estimate is reported. Also, whether the input is balanced or unbalanced, the positive sequence provided by the PMU must be correct at all frequencies. As a practical matter, the PMU Standard calls for this specification to hold over a frequency deviation of $\pm$ 5 Hz from the nominal frequency. Other new features of the standard specify the measurement accuracy requirements for two classes of PMUs, and a standardized reporting time for phasors which is phase-locked to the GPS 1 pps, and is at intervals which are multiples of nominal power frequency periods.

It is also important to note that the standard does not specify the requirements for response of PMUs to power system transients. No doubt this will be covered in forthcoming revisions of the standard. A discussion of transient response of PMUs will be found in Chapter 6.

This section has provided a brief summary of the current PMU standard. The interested reader should refer to the standard document for additional detailed information about requirements for interoperability of PMUs.

5.6.2 File structure of 'Synchrophasor' standard

The file structure for "Synchrophasors" is similar to that of COMTRADE [12], which defines files for transient data collection and dissemination. COMTRADE standard has been adapted by International Electrotechnical Commission (IEC), and is now the principal international file format standard being used by computer relays, digital fault recorders, and other producers and users of power system transient data.

Synchrophasor standard defines four file types for data transmission to and from PMUs. Three files are generated by PMUs: Header files, Configuration files, and Data files. One file, the "Command File", is for communicating with the PMUs from a higher level of the hierarchy – such as a PDC. All files have a common structure as shown in Figure 5.8.

The first word of 2 bytes is for synchronization of the data transfer. The second word defines the size of the total record, the third word identifies the data originator uniquely, and the next two words provide the "second of century" (SOC) and the "fraction of a second" (FRACSEC) at which the data is being reported. The length of the Data words which follow FRACSEC depends upon the specifications provided in the Configuration file. The last word is the check sum to help determine any errors in data transmission.

First
transmitted

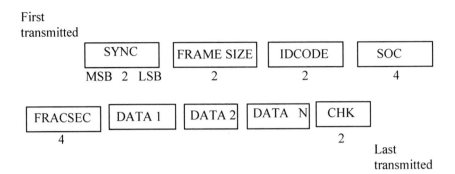

Fig. 5.8 Format for files transmitted from and to PMUs. The numbers below the boxes indicate length of the word in bytes.

The Header file is a human readable file, with pertinent information which the producer of data may wish to share with the user of the data. The Configuration and Data files are machine readable files with fixed formats. Configuration file provides information about the interpretation of the data contained in the data files. In practice the Header and Configuration files are sent by the PMU when the nature of the data being transferred is defined for the first time. The data files contain phasor data (and certain other related measurements such as frequency and rate of change of frequency) which is the principal output of the PMUs. Phasor data may be communicated in rectangular or polar form.

Command files are used by higher levels of the hierarchy for controlling the performance of the PMUs. Several commands have been defined and are available at this time, with a number of reserved codes for commands which may be needed in the future.

Finally, it must be mentioned that this brief summary of the "Synchrophasor" standard is meant to give the reader an idea of how the standard evolved to its present form, and the general structures of its files. One must go to the Standard document for additional details of the file structure and definitions.

5.6.3 PDC files

The PDC and the super PDC (SPDC) of Figure 5.3 are important elements of the overall PMU system organization. Their principal functions are to collate data from different PMUs with identical time-tags, to create archival files of data for future retrieval and use, and to make data stream available to application tasks with appropriate speed and latency. As yet there are no industry standards for the PDC data files. However, it is generally understood that PDCs will have file structures similar to those of PMUs. There are no commercially available PDCs at this time. Most existing PDCs have been custom built by researchers or manufacturers of PMUs. As wider implementation of PMU technology takes place, the industry will no doubt work toward creating standards for these important components of the overall PMU infrastructure.

References

1. Phadke, A.G. and Thorp, J.S., "Computer relaying" (book), Research Studies Press, John Wiley and Sons, New York.
2. "IEEE Standard for Synchrophasors for Power Systems", C37.118-2005. Sponsored by the Power System Relaying Committee of the Power Engineering Society, pp 56–57.
3. Khatib, A.-R.A., "Internet-based Wide Area Measurement Applications in Deregulated Power Systems", Ph.D. Dissertation Virginia Tech, July 2002.
4. Snyder, A.F. et al., "Delayed-Input Wide-Area Stability Control with Synchronized Phasor Measurements and Linear Matrix Inequalities", Power Engineering Society Summer Meeting, 2000. Vol. 2, 16–20 July 2000, pp 1009–1014.
5. Horowitz, S.H. and Phadke, A.G., "Power System Relaying", (book), Third Edition, Research Studies Press Ltd., 2008, Chapter 6.
6. Hecht, J., Understanding Fiber Optics, 4th ed., Prentice-Hall, Upper Saddle River, NJ, USA 2002 (ISBN 0-13-027828-9).
7. "Internet Protocol - DARPA Internet Program Protocol Specification", Postel, J. (ed.), RFC 791, USC/Information Sciences Institute, September 1981.
8. Postel, J.,"Internet Protocol, " RFC 760, USC/Information Sciences Institute, January 1980.
9. Apostolov, A. "Communications in IEC 61850 Based Substation Automation Systems", Power Systems Conference: Advanced Metering, Protection, Control, Communication, and Distributed Resources, 2006. PS '06, 14–17 March 2006, pp 51–56.
10. "IEEE Standard for Synchrophasors for Power Systems", IEEE 1344-1995. Sponsored by the Power System Relaying Committee of the Power Engineering Society.
11. Depablos, J., Centeno, V., Phadke, A.G., and Ingram, M., "Comparative testing of synchronized phasor measurement units", Power Engineering Society General Meeting, 2004. IEEE, Vol. 1, 6–10 June 2004, pp 948–954.
12. "IEEE Standard Common Format for Transient Data Exchange (COMTRADE) for Power Systems", IEEE C37.111-1991, Sponsored by the Power System Relaying Committee of the Power Engineering Society.

Chapter 6 Transient Response of Phasor Measurement Units

6.1 Introduction

As has been mentioned before, phasor is a steady-state concept. In its classical interpretation it is a complex number representation of a pure sinusoidal waveform of single frequency although the frequency is not explicitly displayed in the phasor representation. In power systems, usually the implied frequency is the nominal frequency of the power system. However, it is well known that the power system is rarely operating precisely at the nominal system frequency. We have considered the effect of steady-state off-nominal frequency signals on the phasor measurement process in Chapter 3. It is also well-known that the power system voltages and currents have various harmonic, nonharmonic, and transient components. The harmonic and nonharmonic (steady-state) components are filtered to various degrees by appropriate analog or digital filters.

There are transient phenomena occurring on power systems due to a variety of causes which produce transient components in current and voltage waveforms. A phasor measurement unit (PMU) calculates phasors from sampled data continuously, and it is certain that some of these phasor estimates will involve sampled data containing transient components. The subject of this chapter is to investigate the nature of PMU response to various power system transients.

To consider the transient response of a PMU, we must consider the chain of components in the signal path from the power system up to the phasor output delivered by the PMU. Principal elements of this chain are shown in Figure 6.1. Power system transients result from faults, switching operations, and relative movement of large generator rotors. These sources of transients are represented symbolically in Figure 6.1. In the next section we will consider the nature of transients generated by each of these phenomena. The voltages and currents of the power system are converted to

A.G. Phadke, J.S. Thorp, *Synchronized Phasor Measurements and Their Applications*,
DOI: 10.1007/978-0-387-76537-2_6, © Springer Science+Business Media, LLC 2008

lower-level signals by current and voltage transformers (instrument transformers). The signals are then processed by analog and digital filters serv-
ing the purpose of surge suppression, anti-aliasing filtering, and decimation filtering as appropriate. The filtered signals are then sampled before
phasor computation is performed in the PMU processor. Each of the components identified in Figure 6.1 affects the transient waveforms and therefore the final phasor output produced by the PMU. The following sections
will consider these phenomena in some detail.

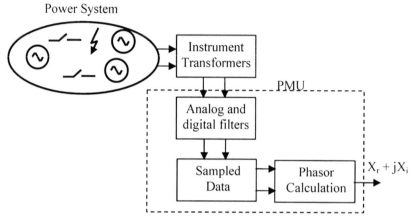

Fig. 6.1 Generation and passage of transient phenomena from power system to the
output of the PMU. Principal components of the chain which affect transient response of the PMU are illustrated.

6.2 Nature of transients in power systems

For the purpose of the present discussion, transients in power systems may
be classified into two categories: electromagnetic transients and electromechanical transients.

6.2.1 Electromagnetic transients

Switching operations and faults produce step changes in the voltage and
current waveforms. On long transmission lines and in transformer and reactor windings there may be multiple reflections of generated transients.
Resonances in the power network create additional frequencies in the
waveforms during these phenomena. Another source of electromagnetic

transients is lightning. Such transients may be classified as electrical or electromagnetic transients [1,2]. The effect of these transients is to introduce high-frequency components in the signals. These transients dissipate within a short time and the waveforms then return to a quasi-steady-state condition. The frequencies of signals produced by electromagnetic transients can be summarized as shown in Figure 6.2.

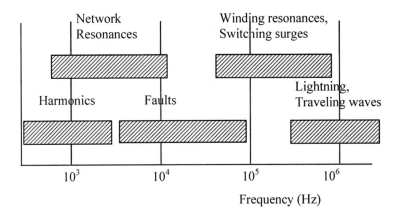

Fig. 6.2 Frequencies of various types of electromagnetic transients. Lightning phenomena have steep wave fronts, which may be interpreted as frequencies as shown above.

The harmonic phenomena indicated in Figure 6.2 are typically produced by power electronic devices. It should be noted that the harmonics frequencies are multiples of the prevailing network frequency, which may be different from the nominal power frequency. (It should be remembered that phasor estimate performed with sampling rates keyed to nominal power frequency do not eliminate harmonics of off-nominal power system frequency.)

Network resonances are caused by shunt and series capacitors, and line charging capacitances interacting with various reactances in the network. Winding resonance phenomena are characteristic of transformer, generator, and reactor windings. Faults may produce some very high-frequency components, especially if arcing is involved in the fault.

With the exception of phenomena at the low end of the spectrum (100–1000 Hz) the electromagnetic transients are severely attenuated by the filters employed in PMUs. Step changes induced in voltages and currents by the faults, and the low-frequency oscillations induced by electromechanical phenomena discussed next are the phenomena of concern in phasor estimation.

6.2.2 Electromechanical transients

Electromechanical transients are created by movement of rotors of large generators (and motors) connected to power networks. Normally all rotors are operating at synchronous speed, with rotor angles at relative positions which are required to meet the load demand on the network. When disturbances such as faults or line outages take place, there is an imbalance between the mechanical and electrical powers of the machines, and rotor angles begin to deviate from their steady-state values. The rotor movement causes the frequency of the generated voltage to deviate from its nominal value. The movement of rotors of different machines occurs at different rates depending upon their moment of inertias and the power imbalance between the inputs and outputs. In general, all generators operate at speeds which may differ from synchronous speed by different (although small) amounts.

Consider the power system shown in Figure 6.3 with "n" generators and "m" buses. Each machine is operating at a different speed, and its output is represented by the current of the appropriate frequency and the generator internal voltage bus is identified as the source of the current. The power system is shown to have only three generators, but in general there will be a much larger number of machines contributing current injections to the network.

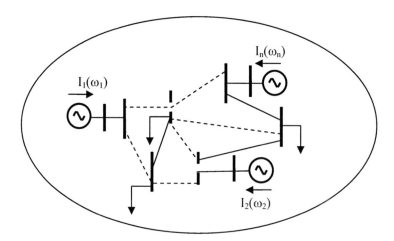

Fig. 6.3 Power system with generators contributing current injections into the network at unequal frequencies.

For the present discussion, we may consider the loads to be represented by impedances. In this case, the above network with "n" generator buses and a total of "m" buses can be represented by a bus impedance matrix $[Z_B]$ of dimension $m \times m$. The voltage at any bus "i" is given by

$$E_i = \sum_{k=1}^{m} Z_{ik} I_k. \tag{6.1}$$

Equation 6.1 clearly shows that the voltage at any network bus is obtained by a superposition of voltages of different frequencies. The contribution of generators which are near bus "i" will be dominant in this superposition.

Superposition of voltages with unequal but close frequencies leads to resultant voltages which show magnitude and phase angle modulation [3]. For example, consider the signal $x(t)$ at a bus obtained by superposition of two signals with frequencies (ω) and ($\omega + \Delta\omega$) with amplitudes X_1 and X_2, respectively. It is assumed that X_2 is much smaller than X_1, implying that the generator which produced X_1 is closer to the bus at which $x(t)$ is being measured. Without loss of generality, we may assume the superimposing signals are sine waves and that their phase angles at $t = 0$ are zero.

$$x(t) = X_1 \sin(\omega t) + X_2 \sin(\omega + \Delta\omega)t \tag{6.2}$$

Using a trigonometric identity Eq. (6.2) can be written as

$$\begin{aligned} x(t) &= X_1 \sin(\omega t) + X_2 \sin(\omega + \Delta\omega)t \\ &= X_3 \sin(\omega t + \phi), \end{aligned}$$

where

$$X_3 = \sqrt{X_1^2 + X_2^2 + 2X_1 X_2 \cos(\Delta\omega t)} \tag{6.3}$$

and

$$\tan(\phi) = \frac{X_2 \sin(\Delta\omega t)}{X_1 + X_2 \cos(\Delta\omega t)}.$$

If X_2 is much smaller than X_1, the expressions for X_3 and ϕ can be simplified:

$$X_3 \cong X_1 + X_2 \cos(\Delta\omega t) \tag{6.4}$$

and

$$\tan(\phi) \cong \frac{X_2}{X_1}\sin(\Delta\omega t) - \frac{1}{2}\left(\frac{X_2}{X_1}\right)^2 \sin(2\Delta\omega t) \cong 0. \qquad (6.5)$$

It is clear from Eq. (6.4) that the effect of this type of superposition is to modulate the amplitude of the signal with a frequency equal to the difference between the two frequencies. The phase angle is also modulated, but for small values of X_2 this effect is negligible.

It is also interesting to note that a pure amplitude modulation of a sinusoid is equivalent to a superposition of three sinusoids of center frequency and two side-bands:

$$\{X_1 + X_2 \cos(\Delta\omega t)\}\sin(\omega t)$$
$$= X_1 \sin(\omega t) + \frac{1}{2}X_2 \sin(\omega + \Delta\omega)t + \frac{1}{2}X_2 \sin(\omega - \Delta\omega)t. \qquad (6.6)$$

Equation 6.6 is an approximation for signal of Eq. (6.2).

Example 6.1　Consider the superposition of two signals:
$$x(t) = 100 \sin(120\pi t) + 10 \sin(100\pi t).$$

The resulting signal is shown in Figure 6.4(a).

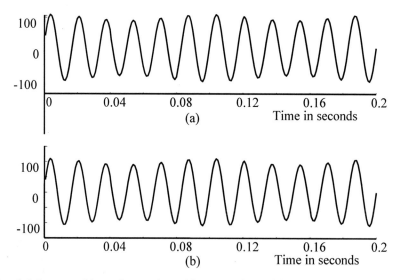

Fig. 6.4 Superposition of two sinusoids approximated by amplitude modulation. (a) Superposition, and (b) Amplitude modulation.

The amplitude modulation is clearly visible in Figure 6.4, and can be approximated by Eq. (6.6):

$$x(t) \cong 100 \ [1+5\cos(20\pi t)] \ \sin(120\pi t).$$

The plot of the approximation is shown in Figure 6.4(b) and is very close to that in Figure 6.4(a).

When several signals of different frequencies are superimposed to form $x(t)$, the resulting expressions are more complex but essentially produce amplitude and phase modulation of the center frequency. This is illustrated in the next example.

Example 6.2 Consider the superposition of three signals:
$$x(t) = 100 \ \sin(120\pi t) + 5 \ \sin(110\pi t) + 10 \ \sin(115\pi t).$$

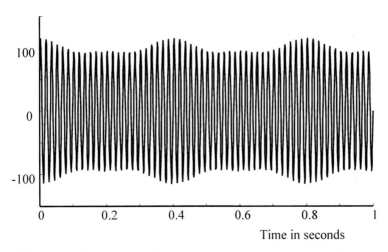

Fig. 6.5 Superposition of three signals representing generators contributions at differing frequencies and magnitudes.

Figure 6.5 is representative of voltage and current waveforms on power networks following large disturbances. The response of PMUs to transients of this nature is one of the most important considerations in determining their ability to be effective in providing feedback to controllers and special protection functions.

6.3 Transient response of instrument transformers

The response of these transformers can be considered under two categories: frequency response and response to step functions. The frequency response is of interest in order to assess how the higher frequencies generated by power system transients will be transmitted to subsequent stages of Figure 6.1 by instrument transformers. As mentioned before, the filtering performed in succeeding stages is likely to suppress most high-frequency components. The step-function response is of particular interest as fault and switching generated transients do contain step-changes in voltages and currents. Saturation effects in current transformers and subsidence effects in CVTs are of particular interest, as they affect the fundamental frequency estimation performed by PMUs.

6.3.1 Voltage transformers

Frequency response

Power systems use two types of voltage transformers: "potential transformers" (PTs) or "capacitive voltage transformers" (CVTs) [4]. Potential transformers are similar to power transformers, with primary and secondary windings on magnetic cores. The PTs have essentially flat response to frequencies which are passed through by the low-pass filters (surge suppression and anti-aliasing filters. Figure 6.6 is adapted from [5] and is representative of frequency response characteristics of several PTs [6]. For the purposes of PMU measurements, we may assume that the PTs have flat response in the frequency range of interest.

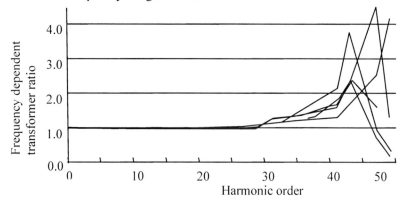

Fig. 6.6 Frequency response characteristics of several potential transformers. Adapted from reference [5].

CVTs have resonances at lower frequencies relative to the PTs. Resonance frequencies encountered in CVT responses depend heavily upon the values of the capacitors, details of the ferro-resonance suppression circuits, and other system components. A representative CVT frequency-dependence response curve is shown in Figure 6.7 [7].

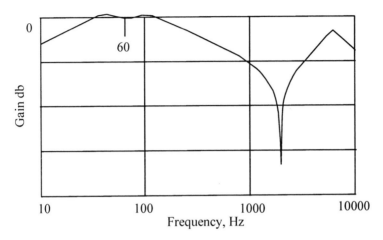

Fig. 6.7 Frequency response characteristics of several capacitive voltage transformers. Adapted from reference [7].

Step function response

For all practical purposes, the step-function response of a potential transformer may be assumed to be transparent. The capacitive voltage transformers have subsidence transient response which depends upon the elements of the voltage transformer circuit, and also upon at what point on the voltage waveform the step function has occurred [4]. Figures 6.8(a) and (b) are representative of CVT subsidence transients.

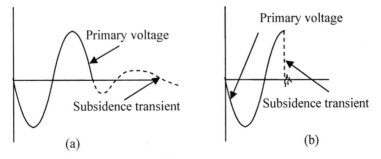

Fig. 6.8 Subsidence transient of a CVT. (**a**) Step function occurring at voltage maximum, and (**b**) Step function occurring at voltage zero.

The subsidence transients are usually associated with faults on the network, and the phasor estimates performed with fault data are usually of no interest in PMU applications. The "transient monitor" function discussed in Section 6.5 below is responsible for identifying the presence of fault data in the data window of PMU estimation, and could be used to flag the phasor estimate as being unusable.

6.3.2 Current transformers

Frequency response

Most current transformers in use in substations are magnetic core multi-winding transformers. Frequency response of these transformers is flat for up to high frequencies, of the order of 50 kHz [8], and will no longer be considered here as the bandwidth of the anti-aliasing filters is much lower than these frequencies.

Step function response

The principal concern in current transformer response to faults is the possibility of saturation of the magnetic core. When the core saturates, the secondary current has severely distorted waveforms depending upon the CT burden, the remanence in the core, and the amount of DC offset in the fault current. A representative CT output waveform due to heavy saturation is shown in Figure 6.9.

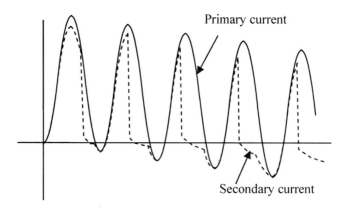

Fig. 6.9 CT saturation. Primary current has a large DC offset, leading to core saturation which may persist for considerable time.

The CT saturation transients are usually associated with fault currents with DC offsets, and the phasor estimates of waveforms during fault conditions are usually of no interest in PMU applications. The "transient monitor" function could be used to identify the presence of fault data in the data window of PMU estimation, and could be used to flag the phasor estimate as being unusable.

6.4 Transient response of filters

6.4.1 Surge suppression filters

Surge suppression filters are designed to block destructive transients created by switching and arcing events in the substations. The frequency of such signals is in the mega-Hertz range [4]. Thus the filters may have cut-off frequencies of the order of hundreds of kiloHertz. It is therefore not necessary to consider the effect of these filters on PMU performance.

6.4.2 Anti-aliasing filters

Frequency response

Anti-aliasing filters used in PMUs have a cut-off frequency which is less than half the sampling frequency. If the signals are sampled at 1000 Hz, the corresponding anti-aliasing filter will have a cut-off frequency of about 400 Hz. The design of anti-aliasing filters may vary considerably depending upon the preference of the PMU manufacturer. We may consider a generic anti-aliasing filter made up of two stages of R-C circuits as shown in Figure 6.10 [9].

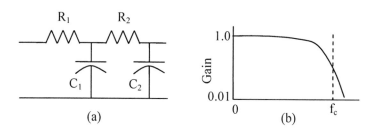

Fig. 6.10 A two-stage R-C anti-aliasing filter. (a) Filter circuit, and (b) Frequency response.

The output of the anti-aliasing filters has a phase lag depending upon the frequency of the input signal. The application functions using PMU data are usually interested in signals with frequencies very close to the nominal power frequency. The phase lag at the output of a two-stage R-C filter with ($R_1 = 1260$ ohms, $R_2 = 2520$ ohms, $C_1 = C_2 = 0.1$ µF) for frequencies in the range of ±5 Hz around the nominal frequency is shown in Figure 6.11. This figure corresponds to an anti-aliasing filter designed to match a sampling rate of 12 times the power system frequency of 60 Hz. It is a simple matter to determine this characteristic for any other filter design. Recall that this phase shift must be compensated in the output of the PMU in accordance with the requirements of the PMU standard discussed in Chapter 5.

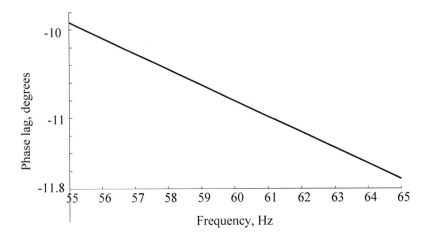

Fig. 6.11 Phase lag produced by a two-stage R-C anti-aliasing filter. The sampling rate is assumed to be 720 Hz, and the filter cut-off frequency is 330 Hz.

Step function response

The anti-aliasing filters respond to step-function inputs caused by faults as shown in Figure 6.12. The filter as in Section 6.4.2.1 is used to produce this figure.

Section 6.5 discusses the response of PMUs to transients caused by faults and other switching events. It will be shown there that phasor estimates produced by PMUs during faults are meant to be discarded, as not representative of the power system state.

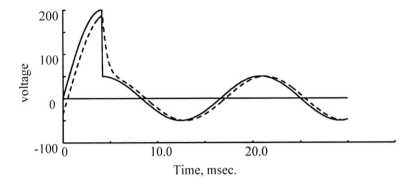

Fig. 6.12 Response of a two-stage R-C anti-aliasing filter of Figure 6.10 to a step change in voltage waveform.

6.5 Transient response during electromagnetic transients

As mentioned in Section 6.4 high-frequency transients will be removed from PMU inputs by the filtering stage. What remains important for our consideration is the effect of step changes in voltages and currents brought about by faults and switching operations. The step changes are further modified by the filtering stage, and the phase angle of the fundamental frequency signal is shifted (lag). This is illustrated in Figure 6.12 for a voltage waveform following a fault which reduces the fundamental frequency voltage to a low value.

Using the recursive phasor estimation process with one cycle data window, the phasor estimate of the pre-fault waveform would be obtained in data windows which contain only the pre-fault data. This is illustrated by data windows 1, 2, and 3 in Figure 6.13 (a) (also see section 2.4). The corresponding phasor is X_1 in Figure 6.13(b). When the data window is fully occupied by post-fault data as with windows N and $N + 1$ in Figure 6.13(a) the phasor estimate becomes X_2 for all succeeding windows. However, while the windows contain both the pre- and post-fault data as with windows 4,5,6,..., the phasor estimate travels along a trajectory from X_1 to X_2 as shown in Figure 6.13(b). These phasor values are not representing the state of the power system, and must be discarded in application of the phasor data.

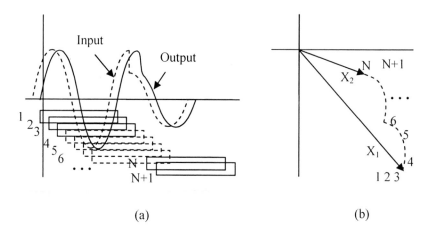

(a)   (b)

Fig. 6.13 Response of PMU to step changes in input signals due to faults or switching operations.

It is important that the transitional phasors be recognized and flagged in order to avoid their use in applications. A possible technique for accomplishing this was described in Section 2.4 with the use of a "transient monitor" function which is a measure of the quality of phasor estimates [9]. This function estimates the difference between the input data samples and the data samples which correspond to the estimated phasor waveform (see Figure 6.14).

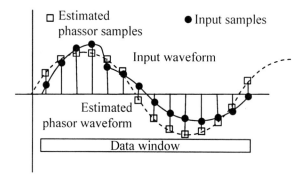

Fig. 6.14 Input data samples and samples corresponding to estimated phasor. The difference between the two is used to define a "transient monitor" function.

Equations (2.26) and (2.27) provide formulas for computing the "transient monitor", and Example 2.4 shows the calculation for a transitional waveform.

The transient monitor (or a similar "quality" function) should be esti-mated at the same time that the phasor is estimated. When this function is outside acceptable bounds, it would be an indication that the phasor esti-mate may be for a transitional signal, or a signal with excessive noise. At present the industry standard does not call for such a specification to be in-cluded in the PMU output. However, it would be a desirable feature which may be provided by manufacturers of PMUs, and eventually adopted in a future standard revision.

6.6 Transient response during power swings

Section 6.2.2 considered the effect of superposition of signals with differ-ing frequencies in power system voltages and currents during transient sta-bility swings. It should be remembered that these swings are relatively slow phenomena, with oscillation frequencies in the range of $0.1-10$ Hz. It has been shown there that superposition effects can be approximated by frequency and amplitude modulation of voltages and currents. The two modulations are likely to be present simultaneously, although it is possible to have amplitude or frequency modulation by itself under certain system conditions. The performance of the synchronized phasor measurement sys-tems for modulated signals is considered in this section. We will consider these modulation phenomena in sequence. In each case the effect of off-nominal frequency operation will be discussed.

6.6.1 Amplitude modulation

Assume that the power system is operating close to nominal frequency (i.e., within ± 5Hz of the nominal frequency), and that a transient stability swing has been initiated. The corresponding voltage and current signals may be assumed to be amplitude-modulated with a frequency ω_m. It is ex-pected that system frequency ω will be close to the nominal power fre-quency of 120π ($\Delta\omega$ varying between -10π and $+10\pi$), while ω_m will vary between $0-20$ π (corresponding to $0-10$ Hz):

$$x(t) = \sqrt{2}[1 + 0.2\sin(\omega_m t)]\cos(\omega t) \tag{6.7}$$

A balanced three-phase input with amplitude modulation as above, and at off-nominal frequency will produce a positive-sequence measurement with just the amplitude modulation.

PMUs estimate positive-sequence phasors continuously, that is, a new phasor is estimated whenever a new data sample is obtained. The data window is equal to one period of the nominal system frequency. Under these conditions, it is clear that the calculated phasors will follow the amplitude modulation perfectly, as long as the modulation frequencies are low.

Example 6.3 A balanced three-phase input having a system frequency excursion of $\Delta\omega = 10\pi$ (5 Hz) and an amplitude modulation frequency of 2Hz is given in Eq. (6.8).

$$x_a(t) = \sqrt{2}[1 + 0.2\sin(4\pi t)]\cos(130\pi t),$$
$$x_b(t) = \sqrt{2}[1 + 0.2\sin(4\pi t)]\cos(130\pi t - 2\pi/3), \qquad (6.8)$$
$$x_c(t) = \sqrt{2}[1 + 0.2\sin(4\pi t)]\cos(130\pi t + 2\pi/3),$$

The result of estimating positive-sequence phasor with a sampling rate of 1440 Hz produces a positive-sequence phasor shown in Figure 6.15 when the amplitude modulation frequency ω_m is set equal to 1, 2, 3, 5, and 10 Hz. It can be seen from this figure that balanced three-phase inputs reproduce amplitude modulation faithfully, and no additional filtering is necessary.

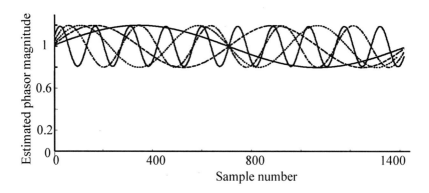

Fig. 6.15 Balanced three-phase inputs at 65 Hz. Positive-sequence magnitude estimate for amplitude modulation at 1, 2, 3, 5, and 10 Hz.

Now consider the case of a single-phase input with amplitude modulation and at off-nominal frequency. It can be expected that in this case the phasor estimate will show a signal at $2(\omega + \omega_0)$ (see Section 3.2), which must be filtered in order to eliminate the possibility of aliasing errors when the phasor output is to be sampled. Figure 6.16 shows the result of estimating the phasor when the input is as given by Eq. (6.7) with $\omega = 65$ Hz, and $\omega_m = 5$ Hz.

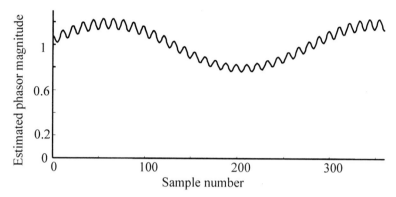

Fig. 6.16 Single-phase input at 65 Hz. Magnitude estimate for amplitude modulation at 5 Hz.

Using the three-point filtering technique of Section 3.3.1, the frequency component at $2(\omega + \omega_0)$ is eliminated for all practical purposes. The result of this filter applied to the phasor estimates in Figure 6.16 is shown in Figure 6.17. It can be seen from these figures that as before, the three-point filter removes the $2(\omega + \omega_0)$ component completely.

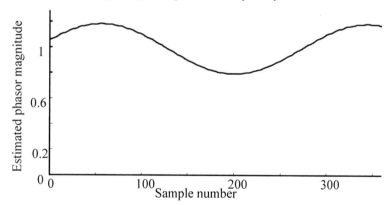

Fig. 6.17 Single-phase input at 65 Hz. Three-point averaging filter. Magnitude estimate for amplitude modulation at 5 Hz.

It should also be noted that in the presence of unbalances, similar remarks about filtering apply for the positive-sequence estimate. When positive-sequence component is estimated, the $2(\omega+\omega_0)$ component is further multiplied by the per unit value of the negative-sequence component, and the three-point filter will remove this component from the positive-sequence estimate.

6.6.2 Frequency modulation

Consider a frequency modulation of the type

$$\omega = \omega_0 + A\sin(\omega_1 t) \tag{6.9}$$

so that the frequency of the input signal varies sinusoidally between $(\omega_0 + A)$ and $(\omega_0 - A)$ with a frequency ω_1. The corresponding phase function ϕ is the integral of the frequency ω:

$$\phi = \omega_0 t - (\frac{A}{\omega_1})\cos(\omega_1 t). \tag{6.10}$$

Balanced three-phase inputs with this phase function are

$$x_a(t) = \sqrt{2}\cos(\phi) = \sqrt{2}\cos\left[\omega_0 t - (\frac{A}{\omega_1})\cos(\omega_1 t)\right],$$

$$x_b(t) = \sqrt{2}\cos(\phi) = \sqrt{2}\cos\left[\omega_0 t - (\frac{A}{\omega_1})\cos(\omega_1 t) - \frac{2\pi}{3}\right], \tag{6.11}$$

$$x_b(t) = \sqrt{2}\cos(\phi) = \sqrt{2}\cos\left[\omega_0 t - (\frac{A}{\omega_1})\cos(\omega_1 t) + \frac{2\pi}{3}\right].$$

Example 6.4

As an example, consider the case of a 60-Hz nominal frequency input varying between 61 Hz and 59 Hz at a frequency of 1 Hz. Since we are considering a single-phase input, we should expect a second harmonic ripple in the estimated phasor angle and magnitude. The result of estimating phasors from a single-phase input are shown in Figure 6.18.

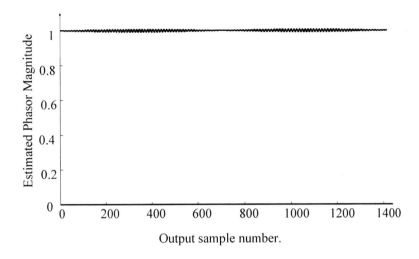

Fig. 6.18 Phasor estimates for single-phase input with frequency modulation.

Note that as the frequency excursions reach extreme values (61 Hz and 59 Hz), the amount of signals at $2(\omega + \omega_0)$ are at a maximum. Note also the 1-Hz modulation frequency reproduced in the envelope of the $2(\omega + \omega_0)$ signal. As before, a three-point averaging filter removes this component almost perfectly. The result is shown in Figure 6.19.

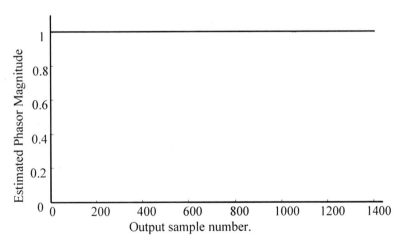

Fig. 6.19 Phasor estimates for single-phase input with frequency modulation. Same case as in Figure 6.18. Effect of a three-point averaging filter.

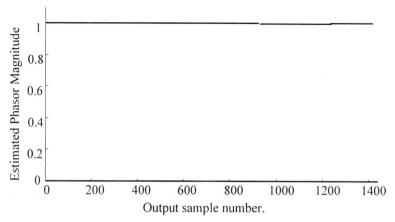

Fig. 6.20 Phasor estimates for positive-sequence component from balanced three-phase inputs with frequency modulation. Frequency excursion and modulation frequency same as in Figure 6.19.

A positive-sequence voltage estimated from a balanced three-phase input is completely free of these $2(\omega + \omega_0)$ components. This is illustrated in Figure 6.20. Although barely noticeable, the amplitude of the estimated phasor does differ from 1.0 as determined by the factor P discussed in Chapters 4 and 5.

6.6.3 Simultaneous amplitude and frequency modulation

During stability swings, both magnitudes and frequency of the three-phase voltages and currents in a power system could be modulated. Equation (6.12) represents such input quantities. Note that the frequency of the amplitude modulation and frequency modulation is equal, namely ω_1.

$$x_a(t) = \sqrt{2}[1+0.2\sin(\omega_1 t)]\cos\left[\omega_0 t - (\frac{A}{\omega_1})\cos(\omega_1 t)\right],$$

$$x_b(t) = \sqrt{2}[1+0.2\sin(\omega_1 t)]\cos\left[\omega_0 t - (\frac{A}{\omega_1})\cos(\omega_1 t) - \frac{2\pi}{3}\right], \quad (6.12)$$

$$x_b(t) = \sqrt{2}[1+0.2\sin(\omega_1 t)]\cos\left[\omega_0 t - (\frac{A}{\omega_1})\cos(\omega_1 t) + \frac{2\pi}{3}\right].$$

The amplitude and frequency modulation both arise out of the superposition of the various generator currents. It is therefore natural that both modulation frequencies be the same.

Example 6.5

The phasor estimate of a single-phase input having a 20% amplitude modulation and a 1-Hz frequency excursion around 60 Hz with both the frequency and amplitude modulation frequency of 2 Hz is shown in Figure 6.21. As can be seen from this figure, the net outcome is a superposition of results in Sections 6.6.1 and 6.6.2.

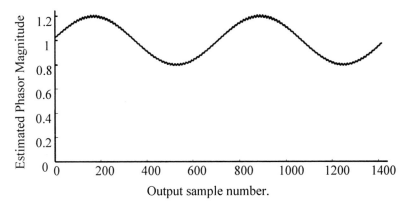

Fig. 6.21 Phasor estimates for single-phase input with frequency and amplitude modulation. Modulation frequency is 2 Hz. Frequency excursions of ± 1 Hz around 60 Hz.

The three-point averaging applied to this estimate effectively eliminates the ripple at $2(\omega + \omega_0)$, as shown in Figure 6.22.

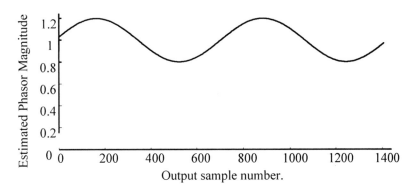

Fig. 6.22 Phasor estimates for single-phase input with frequency and amplitude modulation. Result of applying three-point filter algorithm. Modulation frequency is 2 Hz. Frequency excursions of ± 1 Hz around 60 Hz.

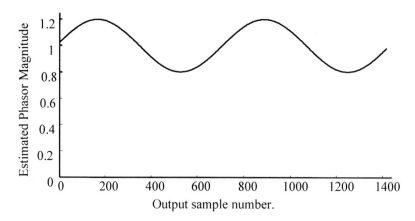

Fig. 6.23 Phasor estimates for balanced three-phase input with frequency and amplitude modulation. Modulation frequency is 2 Hz. Frequency excursions of ±1 Hz around 60 Hz.

As before, balanced three-phase inputs produce positive-sequence phasor estimates which have no ripple at $2(\omega + \omega_0)$ as shown in Figure 6.23. The amplitude modulation in this case was 0.2 per unit zero-to-peak. The output shown in Figure 6.23 (although not discernable from this figure because of the scale used) has a magnitude swing of 0.1995, which is 99.75% of the amplitude of input modulation. The slight drop in measured amplitude modulation is the result of the factor P at off-nominal frequency as explained in Chapters 4–6.

When three-phase inputs are unbalanced, the positive-sequence estimate will once again have a ripple at $2(\omega + \omega_0)$ depending upon the amount of the negative-sequence component present in the input. This ripple can also be eliminated by filtering, the three-point filter being the most efficient one in this regard.

6.6.4 Aliasing considerations in Phasor reporting rates

It is clear from the foregoing discussion that electromechanical transients which reflect the movement of machine rotors are reproduced faithfully by DFT-based phasor estimators. However, since the phasor estimates are reported to higher levels of the hierarchy at a rate which is much lower than that utilized in the figures of this section (once every sample), it becomes necessary to consider the effect of this reporting rate on the frequency re-

sponse observable at the higher levels of hierarchy. Consider a phasor reporting rate of 30 Hz, that is, once every two cycles for a 60-Hz power system. At this reporting rate, the phasors will be able to reproduce correctly
oscillation frequencies lower than 15 Hz without causing errors due to
aliasing. If the power system signals happen to have frequencies higher
than 15 Hz, they must be removed by appropriate filtering before they are
forwarded to applications at the upper hierarchical levels. Table 6.1 sum-
marizes the filtering requirements for various phasor reporting rates.

Table 6.1 Filtering requirements for various phasor reporting rates

Phasor reporting rate (Hz)	Cut-off frequency of phasor processing filters (Hz)
60	30
30	15
20	10
15	7.5
10	5

It should be noted that most electromechanical oscillation frequencies
observed in large power networks are well above the lowest cut-off frequency in Table 6.1. Nevertheless, it is possible that there are small signals
at higher frequencies caused by resonances or by machines of low inertia.
It is therefore necessary to include such a filter in all phasor measurement
and application systems.

Any of the standard filter design techniques could be used to accomplish
this task. A very simple averaging filter with appropriate averaging window may be used as illustrated in the following example.

Example 6.6
Consider a phasor estimate in a 60-Hz power system with 5 Hz and 20
Hz modulating frequencies. The sampling rate is assumed to be 1440 Hz
(i.e., 24 samples per cycle), and the modulating 5 Hz and 20 Hz components are 10% and 20%, respectively, of the fundamental frequency. Considering only the magnitude modulation here, the phasor magnitude of the
*k*th sample may be represented by

$$|X_k| = 1.0 + 0.1\cos(\frac{k}{24 \times 12}) + 0.2\cos(\frac{k}{24 \times 3}).$$

The phasors are to be reported at a rate of 20 Hz, so that all frequency
components 10 Hz and above must be eliminated by the filter. A 10-Hz
signal will have a period of 144 samples (1440/10). Thus, we may

construct an averaging filter with a width of 144 samples. Thus, a new phasor magnitude is created by averaging successive 144 samples of the original data set (see Figure 6.24).

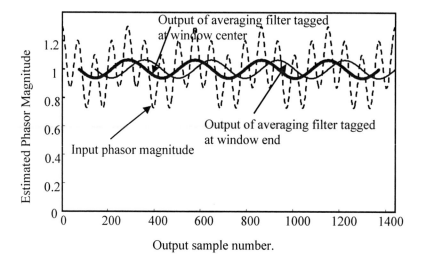

Fig. 6.24 Averaging filter for eliminating frequencies above the Nyquist rate.

It should be noted that the result of averaging is available at the end of the averaging period, which produces the plot with the thin solid line. On the other hand, by assigning the average to the middle of the averaging window, the bold solid line is produced, which has the correct phase information about the modulating signal at 5 Hz.

Signals with magnitude and phase angle modulation are treated similarly. For most practical purposes a simple averaging filter (the so-called box-car filter) is sufficiently accurate to meet the Nyquist criterion associated with the phasor reporting rate.

In summary, the present industry standard (C37.118) for PMUs does not specify the requirements for transient response of PMUs. However, it is to be expected that future revision of the standard will deal with this requirement in order to achieve interoperability between PMUs manufactured by different manufacturers. The principal task of PMUs is to measure positive-sequence voltages and currents in the power network. By definition phasors and positive sequence are steady-state concepts. Thus, as discussed in this chapter phasors estimated during fault and other switching events should be flagged as not representing any state of the network. The swings of machine rotors are treated as a sequence of steady-state conditions, and PMUs are shown to measure the rotor swings accurately for all practical power systems. Finally, the issue of phasor reporting rate and possible aliasing effects due to the presence of higher swing frequencies in the network has also been addressed. It is shown that a simple averaging filter does an effective job of taking care of the aliasing problem, when the window of the averaging filter is adjusted to the phasor reporting rate.

References

1. Greenwood, A., "Electrical transients in power systems", (book) Wiley Interscience, 1971.
2. Chowdhuri, P., "Electromagnetic transients in power systems", (book) Research Studies Press Ltd., John Wiley & Sons, Inc., 1996.
3. Phadke, A.G., Kasztenny, B., "Synchronized phasor and frequency measurement under transient conditions", Accepted for publication in IEEE Transactions on Power Delivery, 2008.
4. Horowitz, S.H. and Phadke, A.G., "Power system relaying", (book), John Wiley & Sons, Second Edition, 4th printing,
5. Seljeseth, H. et al., Voltage transformer frequency response. Measuring harmonics in Norwegian 300 kV and 132 kV Power Systems", 8th International Conference on Harmonics and Quality of Power, ICHQP '98, IEEE/PES and NTUA, Athens, Greece, October 14–16, 1998.
6. Meliopoulos, A.P.S. et al., "Transmission level instrument transformers and transient event recorders characterization for harmonic measurements", IEEE Transactions on Power Delivery, Vol. 8, No. 3, July 1993.
7. Working Group C-5 of IEEE Power System Relaying Committee, "Mathematical models for current, voltage, and coupling capacitor voltage transformers", IEEE Transactions on Power Delivery, Vol. 15, No. 1, January 2000.
8. Samesima, M.I. et al., "Frequency response analysis and modeling of measurement transformers under distorted current and voltage supply", IEEE Transactions on Power Delivery, Vol. 6, No. 4, pp 1762–1768, October 1991.
9. Phadke, A.G. and Thorp, J.S., "Computer relaying for power systems, (book), Research Studies Press, Reprinted May, 2000, p 150.

Part II: Phasor Measurement Applications

Chapter 7 State Estimation

7.1 History – Operator's load flow

Before the advent of state estimation the power system operator had responsibility for many real-time control center functions, including scheduling generation and interchange, monitoring outages and scheduling alternatives, supervising scheduled outages, scheduling frequency and time corrections, coordinating bias settings, and emergency restoration of the system. All of this was done with operating guides produced by the planning department after running a large number of load flows. As is always the case, the actual events faced by the operator were occasionally unexpected and had not been included in the planning cases. The solution was to supply the operator with a load flow program installed in the control center. The operator could manually enter data describing the current situation and get a load flow corresponding to the real world. Unexpectedly, the operator's load flow did not work well. The problems were caused by insufficient data, nonuniform data, and errors in the data and in the model.

It was important that the operator's load flow be accurate so that the outage relief would be accurate. It was also recognized that the planning load flow was not exactly what the operator needed. What was required in the control center was a process of using a large number of imprecise measurements to estimate the existing state of the system. Early state estimation algorithms [1] used measurements of line flows, both real and reactive power, to estimate the bus voltage angles and magnitudes. The complex bus voltages are the state of the system since given an accurate model of the network the bus voltages determine the complex power flows in the lines and all the complex power injections. Unfortunately prior to synchronized phasor measurements, the state could not be measured directly but only inferred from the unsynchronized power flow measurements. This fact and the process of getting large numbers of measurements into the control center forced the first state estimators to make compromises which

A.G. Phadke, J.S. Thorp, *Synchronized Phasor Measurements and Their Applications*, DOI: 10.1007/978-0-387-76537-2_7, © Springer Science+Business Media, LLC 2008

persist today and have an influence on how phasor measurements must be integrated into existing state estimation algorithms.

7.2 Weighted least square

7.2.1 Least square

The terms least squares and weighted least squares have to do with the criterion for selecting a solution to the overdefined equations produced when more measurements than states are represented in the estimation problem. Suppose the equations are linear and in the form

$$\mathbf{y} = \mathbf{A}\mathbf{x}, \tag{7.1}$$

where $\mathbf{x}$ is the state, $\mathbf{y}$ the measurements, and $\mathbf{A}$ is a matrix with more rows than columns. In general Eq. (7.1) does not have a solution and should be written as

$$\mathbf{y} = \mathbf{A}\mathbf{x} + \boldsymbol{\varepsilon}, \tag{7.2}$$

where the vector $\boldsymbol{\varepsilon}$ represents errors in the measurements. The "least squares" solution is based on assuming the errors are independent and identically distributed with mean zero and variance 1; that is, with the unit matrix denoted by $\mathbf{I}$:

$$E\{\boldsymbol{\varepsilon}\} = 0, \quad E\{\boldsymbol{\varepsilon}\boldsymbol{\varepsilon}^T) = \mathbf{I}. \tag{7.3}$$

The optimization problem is to find the estimate, $\hat{\mathbf{x}}$ which minimizes

$$E\{(\mathbf{y} - \mathbf{A}\hat{\mathbf{x}})^T (\mathbf{y} - \mathbf{A}\hat{\mathbf{x}})\} = \mathbf{y}^T\mathbf{y} - 2\mathbf{y}^T\mathbf{A}\hat{\mathbf{x}} - \hat{\mathbf{x}}^T\mathbf{A}^T\mathbf{A}\hat{\mathbf{x}}. \tag{7.4}$$

The solution is

$$\hat{\mathbf{x}} = (\mathbf{A}^T\mathbf{A})^{-1}\mathbf{A}^T\mathbf{y}. \tag{7.5}$$

Equation (7.5) is what Matlab produces for the \ operation; that is, $\hat{\mathbf{x}} = \mathbf{A} \setminus \mathbf{y}$.

Example 7.1

Let
$$\mathbf{A} = \begin{bmatrix} 7 & 2 & 1 \\ 3 & 5 & 0 \\ 1 & 4 & 6 \\ -1 & 3 & 4 \\ 2 & -3 & 5 \end{bmatrix} \quad \mathbf{y} = \begin{bmatrix} 15 \\ 0 \\ 11 \\ 2 \\ 16 \end{bmatrix}.$$

Then

$$\mathbf{A}^{\mathrm{T}}\mathbf{A} = \begin{bmatrix} 64 & 24 & 19 \\ 24 & 63 & 23 \\ 19 & 23 & 78 \end{bmatrix} \quad \mathbf{A}^{\mathrm{T}}\mathbf{y} = \begin{bmatrix} 146 \\ 32 \\ 169 \end{bmatrix}$$

and

$$\hat{\mathbf{x}} = \begin{bmatrix} 2.0744 \\ -0.9961 \\ 1.9551 \end{bmatrix} \quad \mathbf{y} - \mathbf{A}\hat{\mathbf{x}} = \begin{bmatrix} 0.5165 \\ -1.2428 \\ 1.1794 \\ -0.7578 \\ -0.9123 \end{bmatrix}.$$

7.2.2 Linear weighted least squares

Least squares

Suppose there was more information about the errors ε in Eq. (7.2) in the form of a covariance matrix

$$E\{\varepsilon\varepsilon^{\mathrm{T}}\} = \mathbf{W}. \tag{7.6}$$

Even in the simple case where $\mathbf{W}$ is diagonal Eq. (7.6) allows different errors to be weighted differently. The objective is to minimize

$$E\{(\mathbf{y} - \mathbf{A}\hat{\mathbf{x}})^T \mathbf{W}^{-1}(\mathbf{y} - \mathbf{A}\hat{\mathbf{x}})\} =$$
$$\mathbf{y}^T\mathbf{W}^{-1}\mathbf{y} - 2\mathbf{y}^T\mathbf{W}^{-1}\mathbf{A}\hat{\mathbf{x}} - \hat{\mathbf{x}}^T\mathbf{A}^T\mathbf{W}^{-1}\mathbf{A}\hat{\mathbf{x}}, \qquad (7.7)$$

which has solution

$$\hat{\mathbf{x}} = (\mathbf{A}^T\mathbf{W}^{-1}\mathbf{A})^{-1}\mathbf{A}^T\mathbf{W}^{-1}\mathbf{y}. \qquad (7.8)$$

An understanding of weighted least squares can be obtained by considering the diagonal version of the matrix $\mathbf{W}$. If $\mathbf{W}$ is diagonal then the objective function is simply

$$E\{(\mathbf{y} - \mathbf{A}\hat{\mathbf{x}})^T \mathbf{W}^{-1}(\mathbf{y} - \mathbf{A}\hat{\mathbf{x}})\} = \sum \frac{(y_i - \hat{y}_i)^2}{W_{ii}}, \qquad (7.9)$$

where

$$\hat{\mathbf{y}} = \mathbf{A}\hat{\mathbf{x}} \quad \tilde{\mathbf{y}} = \mathbf{y} - \hat{\mathbf{y}}. \qquad (7.10)$$

The variable $\tilde{\mathbf{y}}$, the difference between the actual measurement and the estimated measurement in Eq. (7.9) or (7.4) is called the measurement residual. When the measurement errors are identically distributed we are minimizing the sum of the squares of these residuals. When the measurement errors are independent but of different sizes we are dividing the residuals by the measurement variances to normalize things; that is, if a measurement error is large then a larger residual is accepted while a small measurement error demands a smaller residual. The mean of $\tilde{\mathbf{y}}$ is zero and the covariance of $\tilde{\mathbf{y}}$ is

$$E\{\tilde{\mathbf{y}}\} = 0 \quad Cov(\tilde{\mathbf{y}}) = \mathbf{A}(\mathbf{A}^T\mathbf{W}^{-1}\mathbf{A})^{-1}\mathbf{A}^T. \qquad (7.11)$$

A tall $\mathbf{A}$ matrix can cause numerical difficulties. State estimation frequently involves thousands of measurements. $\mathbf{A}$ could have thousands of rows and in that case the QR algorithm can be employed. If we write

$$\mathbf{W}^{-1} = \mathbf{M}^T\mathbf{M}, \quad \mathbf{M}\mathbf{A} = \mathbf{Q}\begin{bmatrix} \mathbf{R} \\ \mathbf{0} \end{bmatrix}, \qquad (7.12a)$$

where $\mathbf{Q}$ is orthogonal, that is, $\mathbf{QQ}^T = \mathbf{I}$ and R is upper triangular then

$$(\mathbf{A}^T\mathbf{W}^{-1}\mathbf{A})^{-1}\mathbf{A}^T\mathbf{W}^{-1} = \mathbf{R}^{-1}[\mathbf{I} \quad \mathbf{0}]\mathbf{Q}^T\mathbf{M}. \tag{7.12b}$$

Example **7.2**

Suppose everything is the same as in Example 7.1 except

$$\mathbf{W} = \begin{bmatrix} 1 & 0 & 0 & 0 & 0 \\ 0 & 1 & 0 & 0 & 0 \\ 0 & 0 & 1 & 0 & 0 \\ 0 & 0 & 0 & 100 & 0 \\ 0 & 0 & 0 & 0 & 100 \end{bmatrix},$$

that is, the last two measurements are 10 times larger than the first three,

$$\hat{\mathbf{x}} = \begin{bmatrix} 2.1762 \\ -1.2946 \\ 2.3277 \end{bmatrix} \qquad \tilde{\mathbf{y}} = \begin{bmatrix} 0.0280 \\ -0.0557 \\ 0.0360 \\ -1.2508 \\ -3.8746 \end{bmatrix}.$$

The residuals corresponding to the better measurements are considerably smaller than the residuals for the poor measurements as would be expected. The covariance matrix for the measurement residual also shows the effect of the unequal measurement error variances. The first 3×3 block of covariance matrix is almost a unit matrix while the last two residuals are strongly correlated with the first three.

$$\mathrm{Cov}(\tilde{\mathbf{y}}) = \begin{bmatrix} 0.9925 & 0.0123 & -0.0033 & -0.3076 & 0.7964 \\ 0.0123 & 0.9784 & 0.0089 & 0.1486 & -1.3378 \\ -0.0033 & 0.0089 & 0.9902 & 0.7136 & 0.6733 \\ -0.3076 & 0.1486 & 0.7136 & 0.6298 & 0.0227 \\ 0.7964 & -1.3378 & 0.6733 & 0.0227 & 3.2611 \end{bmatrix}$$

7.2.3 Nonlinear weighted least squares

If the measurements are a nonlinear function of the state

$$z = h(x) + \varepsilon, \quad E(\varepsilon) = 0, \quad E(\varepsilon\varepsilon^T) = W, \tag{7.13}$$

then the task is to find $\hat{x}$ to minimize

$$J(\hat{x}) = [z - h(\hat{x})]^T W^{-1}[z - h(\hat{x})]. \tag{7.14}$$

Equation (7.14) must be minimized recursively by linearizing $h(x)$ about x^k, the value of x at the last iteration,

$$h(x) = h(x^k) + H(x - x^k), \tag{7.15}$$

where H is a matrix of first partial derivatives of the elements of h with respect to the components of x evaluated at x^k. If

$$\Delta\hat{x} = x - x^k, \quad \Delta z = z - h(x^k), \tag{7.16}$$

then one step in the iteration is given by the solution of Eq. (7.18) which is an incremental version of Eq. (7.8):

$$H^T W^{-1} H \, \Delta x = H^T W^{-1} \Delta z, \tag{7.17}$$

$$\Delta x = \left(H^T W^{-1} H\right)^{-1} H^T W^{-1} \Delta z. \tag{7.18}$$

The covariance of the resulting estimate is the given by the matrix $(H(\hat{x})W^{-1}H(\hat{x}))^{-1}$.

Example 7.3

Let

$$h(x) = \begin{bmatrix} x^2 - 4x + 4 \\ 3x - 1 \end{bmatrix} \quad z = \begin{bmatrix} 4 \\ 1 \end{bmatrix} \quad H = \begin{bmatrix} 2x - 4 \\ 3 \end{bmatrix} \quad W = \begin{bmatrix} 1 & 0 \\ 0 & 1 \end{bmatrix}.$$

From Eq. 7.18 the iteration is

$$x^{k+1} = x^k + \frac{1}{(2x^k - 4)^2 + 9}\begin{bmatrix} 2x^k - 4 & 3 \end{bmatrix}\begin{bmatrix} 4 - (x^k)^2 + 4x^k - 4 \\ 1 - 3x^k + 1 \end{bmatrix}.$$

Starting at $x^0 = 4$ the values are given in Table 7.1

Table 7.1 The first seven iterations for Example 7.3

k	x^k
1	2.80
2	1.6042
3	0.4416
4	0.2861
5	0.2756
6	0.2746
7	0.2745

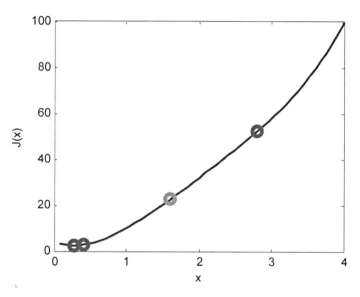

Fig. 7.1 The objective function and the first four iterations.

7.3 Static state estimation

The power system application of the preceding begun after the 1965 Northeast blackout involved analog measurements and less sophisticated communication systems than available today. These facts forced an approximation on the problem from the outset. The measurements from a supervisory control and data acquisition (SCADA) system composed of remote terminal units (RTUs) in the substations were obtained in sequence by polling. The data scan took long enough that the system was actually in a slightly different state when the scan was complete than it had been at the beginning. The approximation was to assume the system did not change during the scan – that the system was "static". Perhaps it is better to view the resulting estimate of the state to be the state of a hypothetical system which could simultaneously support the complete set of measurements. Of course, depending on how long the scan takes and what changes in load and generation take place during the scan, the hypothetical system may not exist or may be quite different from the real system. While scans have become quicker the introduction of phasor measurements forces reconsideration of the static assumption.

The state of the system is the collection of system bus voltages which are complex numbers. Conventional static state estimation algorithms use the magnitude and angle of these voltages as the states and measurements of real and reactive power flows and injections along with some voltage magnitude measurements. The use of synchronized phasor measurements can make it attractive to also consider the problem in rectangular coordinates with real and imaginary parts of voltages and currents.

Elaborate error models were developed [2] for the analog measurements that had dependence on the size of the measurement and the full scale of the meter. Digital technology makes these models less appropriate. The common elements in measurements, both old and new, are the current transformers, CTs, and voltage transformers, PTs. The CTs, in particular, have a fixed bias that is not modeled with the measurement error ε described in the section on weighted least-squares.

In polar coordinates the angles θ_p and magnitudes V_p are the states where

$$E_p = V_p e^{j\theta_p}. \tag{7.19}$$

The complex flows and injections are nonlinear. For example, the real power flow from bus p to bus q with a series impedance z_{pq} between p and q and a shunt admittance at bus p of y_p is

$$P_{pq} = V^2{}_p [|Y_{pq}|\cos(\beta_{pq}) + |y_p|\cos(\alpha_p)$$
$$- V_p V_q \cos(\theta_p - \theta_q - \beta_{pq}), \tag{7.20}$$

where α_p and β_{pq} are the angles in Eq. (7.21):

$$\frac{1}{z_{pq}} = \left| Y_{pq} \right| e^{-j\beta_{pq}} \qquad y_p = \left| y_p \right| e^{j\alpha_p}. \tag{7.21}$$

In general

$$\mathbf{z} = \mathbf{h}(\mathbf{V}, \boldsymbol{\theta}) + \boldsymbol{\varepsilon}. \tag{7.22}$$

With assumptions as in Section 7.2 that the measurement errors have zero mean and are independent,

$$E\{\boldsymbol{\varepsilon}\} = 0, \quad E\{\boldsymbol{\varepsilon}\boldsymbol{\varepsilon}^T\} = \mathbf{W}, \quad w_{ij} = 0, \quad w_{ii} = \sigma_i^2. \tag{7.23}$$

The estimate is formed by minimizing

$$J(\mathbf{V}, \boldsymbol{\theta}) = [\mathbf{z} - \mathbf{h}(\mathbf{V}, \boldsymbol{\theta})]^T \mathbf{W}^{-1} [\mathbf{z} - \mathbf{h}(\mathbf{V}, \boldsymbol{\theta})]$$

$$= \sum_{i=1}^{m} \frac{(z_i - h_i(V, \theta))^2}{\sigma_i^2}. \tag{7.24}$$

Following the steps in Eqs. (7.16–7.18)

$$\mathbf{h}(\mathbf{V}, \boldsymbol{\theta}) = \mathbf{h}(\mathbf{V}^k, \boldsymbol{\theta}^k) + \mathbf{H} \begin{bmatrix} \mathbf{V} - \mathbf{V}^k \\ \boldsymbol{\theta} - \boldsymbol{\theta}^k \end{bmatrix}, \tag{7.25}$$

where $\mathbf{H}$ is a matrix of first partial derivatives of the elements of $\mathbf{h}$ with respect to the components of $\mathbf{x}$ evaluated at x^k. If

$$\begin{bmatrix} \Delta\mathbf{V} \\ \Delta\boldsymbol{\theta} \end{bmatrix} = \begin{bmatrix} \mathbf{V} - \mathbf{V}^k \\ \boldsymbol{\theta} - \boldsymbol{\theta}^k \end{bmatrix}, \quad \Delta\mathbf{z} = \mathbf{z} - \mathbf{h}(\mathbf{V}^k, \boldsymbol{\theta}^k), \tag{7.26}$$

then one step in the iteration is given by

$$\mathbf{H}^T \mathbf{W}^{-1} \mathbf{H} \begin{bmatrix} \Delta\mathbf{V} \\ \Delta\boldsymbol{\theta} \end{bmatrix} = \mathbf{H}^T \mathbf{W}^{-1} \Delta\mathbf{z},$$

$$\mathbf{G} \begin{bmatrix} \Delta\mathbf{V} \\ \Delta\boldsymbol{\theta} \end{bmatrix} = \mathbf{H}^T \mathbf{W}^{-1} \Delta\mathbf{z}. \tag{7.27}$$

The gain matrix $\mathbf{G}$ in Eq. (7.27) is large and sparse and Eq. (7.27) is solved with Gaussian elimination. Note that $\mathbf{H}$ is much like the load flow Jacobian in terms of sparsity. With organization into active and reactive power the equivalent of the fast decoupled load flow can be used to simplify Eq. (7.27). First order the measurements into real or active power (sub A) and reactive power (sub R)

$$\mathbf{z} = \begin{bmatrix} \mathbf{z}_A \\ \mathbf{z}_R \end{bmatrix}, \quad \mathbf{z}_A = \begin{bmatrix} P_{km} \\ P_k \end{bmatrix}, \quad \mathbf{z}_R = \begin{bmatrix} Q_{km} \\ Q_k \\ V_k \end{bmatrix}, \tag{7.28}$$

and then write the states as angles followed by voltage magnitudes to form

$$\begin{bmatrix} \mathbf{G}_{AA} & 0 \\ 0 & \mathbf{G}_{RR} \end{bmatrix} \begin{bmatrix} \Delta\theta \\ \Delta v \end{bmatrix} = \mathbf{H}^T \mathbf{W}^{-1} \begin{bmatrix} \Delta\mathbf{z}_A \\ \Delta\mathbf{z}_R \end{bmatrix}. \tag{7.29}$$

The off-diagonal blocks in Eq. (7.29) are zero under the same assumptions that are used in the fast decoupled load flow, namely that angle differences are small, voltage magnitudes are near 1, and that the X/R of the lines are large. An even stronger assumption can be made and G computed with angles set equal to zero and voltage magnitudes set equal to 1. Note in this case **G** does not have to be recomputed between iterations. The price is inevitably that more iterations are required.

Example 7.4

A 30-bus system is shown in Figure 7.2. Figure 7.3 shows the errors in bus voltage angles and magnitudes for a specific case. All real and imaginary flows and injections are measured with a random error with a sigma of 1% of the magnitude of the complex power and all voltage magnitudes are measured with a sigma of 1% per unit. The results correspond to a specific set of random errors added to a load flow solution.

The data for the 30-bus system along with state estimation program used here is included in a suite of free software available at the matpower website http://www.pserc.cornell.edu/matpower. The program is contained in a folder called extras\state estimation. It can be used on a variety of cases available in matpower. The function state_est.m has bus numbering specific to bus 1 being the swing bus connected to buses 2 and 3 as in Figure 7.2.

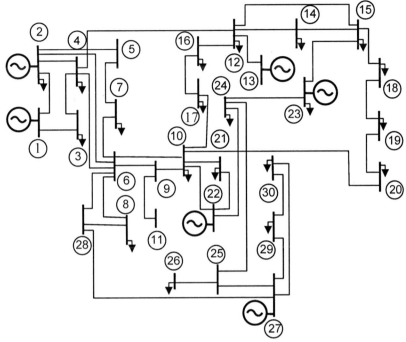

Fig. 7.2 Thirty-bus system for Example 7.4.

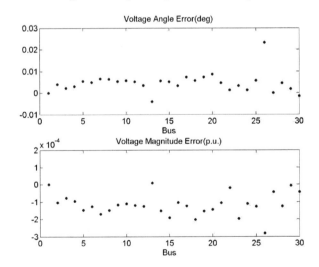

Fig. 7.3 Bus voltage angle and magnitude errors for the 30-bus system in Figure 7.2.

7.4 Bad data detection

One of the most important functions of a state estimator is to identify and reject bad data [3]. Bad data can arise from problems in the measuring unit or in the communication of that data. If it is caused by an uncalibrated measuring instrument it probably is modest in size and may even fit within the model of the random errors in Eq. (7.22). A communication error however might produce an immense error. The estimator could be seriously damaged by one huge measurement error. The solution, of course, is to eliminate measurements that have such large errors before performing the calculations. This is possible because of the ability to compute the measurement residuals.

$$\tilde{\mathbf{z}} = [\mathbf{z} - \mathbf{h}(\hat{\mathbf{V}}, \hat{\mathbf{\theta}})] . \tag{7.30}$$

Equation (7.11) gave the covariance of $\tilde{\mathbf{y}}$ in the linear case. Recognizing the connection between $\mathbf{A}$ in the linear case and $\mathbf{H}$ in the nonlinear case the covariance of $\tilde{\mathbf{y}}$ is given by

$$\mathbf{R} = \text{Cov}(\tilde{\mathbf{y}}) = \mathbf{H}(\mathbf{H}^T \mathbf{W}^{-1} \mathbf{H})^{-1} \mathbf{H}^T . \tag{7.31}$$

If we normalize the vector of residuals by their covariance matrix

$$c = \tilde{\mathbf{y}}^T \mathbf{R}^{-1} \tilde{\mathbf{y}} , \tag{7.32}$$

we obtain a χ^2 (chi-squared) random variable. It has $E\{c\} = m$, the number of measurements and is concentrated around its mean. The probability density for $m = 40$ is shown in Figure 7.4

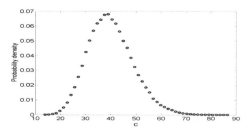

Fig. 7.4 Chi-squared density with 40 degrees of freedom.

The density become more concentrated as the number of degrees of freedom (m) increases so that with some confidence the quantity c can be used to determine if some data fits the model. For $m = 40$ in the figure if c was greater than 60 or less than 20 it would be reasonable to say something was wrong. For any m upper and lower bounds on c can be set to initiate further tests of individual residuals.

The individual residuals can be normalized by their variance. One procedure, largest normalized residual (LNR), test is as follows:

1. Normalize the residuals by the measurement variances

$$\tilde{z}_{in} = [z_i - h_i(\hat{V}, \hat{\theta})]/\sigma_i \qquad (7.33)$$

where the subscript n indicates a normalized quantity.

2. Rank the normalized residuals.

3. Eliminate the measurements with residuals above some threshold or simply the largest.

4. Repeat the estimation problem without the measurements in 3.

5. Check c again.

6. Return to 1 if c is too large.

The m functions used in *Example* 7.4 also use LNR and rejects data with a normalized residual greater than 2.5. In fact the case shown in Figure 7.3 rejected a single measurement.

To show the effect of interacting bad data consider the system in Figure 7.2.

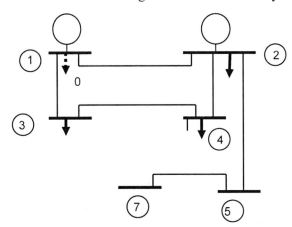

Fig. 7.5 A portion of the system in Figure 7.3 with one bad measurement.

The affected section is redrawn in Figure 7.5. If the measurement of the real flow from bus 3 to bus 4 is zero (the actual flow is 12.56 MW) the bad data detection identifies it and rejects it. The results of the estimate after rejecting the bad data is shown in Figure 7.6

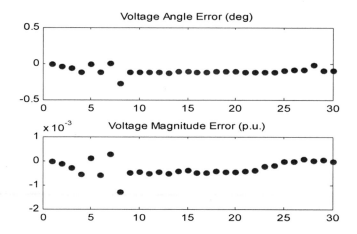

Fig. 7.6 The errors with one bad measurement of the real power flow from bus 3 to 4 removed.

If interacting bad data is added in the form of another zero measurement of real flow from bus 3 to bus 4 then the bad data rejection ranks the flow from bus 2 to bus 1 as the first bad data, rejects it and then rejects the reactive injection at bus 7 in the next iteration. The resulting errors are shown in Figure 7.7. Larger errors in both voltage angle and magnitude in the first eight buses are visible.

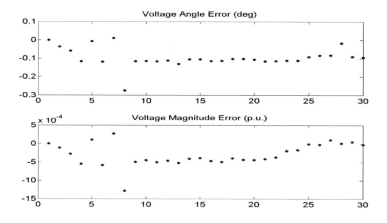

Fig. 7.7 The errors with interacting data. The measured flow from 1 to 2 was incorrectly identified as bad data along with the reactive injection at bus 7.

The issue recognized in [4] is that bad data can reinforce itself and force the LNR procedure to eliminate good data. If the bad data is statistically independent the interaction is unlikely.

7.5 State estimation with Phasors measurements

The addition of even a few direct measurements of angle to the previous formulation has a number of advantages and creates a symmetry in the problem statement. Equation (7.28) becomes Eq. (7.34), when angle measurements are included.

$$z = \begin{bmatrix} z_A \\ z_R \end{bmatrix}, \quad z_A = \begin{bmatrix} P_{km} \\ P_k \\ \theta_k \end{bmatrix}, \quad z_R = \begin{bmatrix} Q_{km} \\ Q_k \\ V_k \end{bmatrix}. \tag{7.34}$$

The matrix $\mathbf{H}$ is modified in an obvious manner but otherwise the previous development applies. The matpower m files from Example 7.4 have the angle measurements included. If the structure of Example 7.4 is maintained and a complete set of phasor angle measurements are added with measurement errors variances of 0.02 degrees the performance shown in Figure 7.8 results.

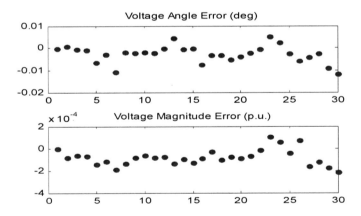

Fig. 7.8 The errors with a complete set of phasor measurements of angles with a sigma of 0.02 degrees.

Note the angle errors are smaller as would be expected. A complete set of angle measurements would be quite expensive and is used only as an example. We will consider the selection of the location for a few phasor measurements in the sequel. One issue that must be remembered is that phasor measurements have universal time as a reference, that is, the sampling instants determine the reference for the PMU data. The conventional state estimator has a particular bus as a reference. If the angle measurements are added without considering the different references the algorithm is liable not to converge. The solution is to obtain a common reference. An obvious approach is to measure the angle of the bus that is the reference for the conventional estimator with a PMU.

In addition to measuring bus voltages, PMUs can measure the currents in lines connected to the bus. The addition of this data further complicates the formulation because it creates a tension between rectangular and polar coordinates. The preceding is a polar formulation with the PMU measurement modeled as a measurement of voltage angle. The actual measurement is inherently one of the real and imaginary parts of the bus voltage and line currents. In the next section a linear, rectangular, estimator will be formulated using only these linear PMU measurements. However, integrating line current measurements into a conventional estimator with the systems state expressed in polar coordinates means expressing the line currents as nonlinear functions of the magnitude and angle of the bus voltages or arguing that the PMU measures the magnitude and angle of the line currents (which are still nonlinear functions of the system state). Of course, the angle and magnitude can be computed from the rectangular parts but the issue is the covariance of the measurement errors and the resulting covariance of the error is the estimates.

7.5.1 Linear state estimation

If an estimate could be formed with only PMU data then the issues of data scan and time skew could be eliminated. The PMU data would be time-tagged and the static assumption removed. We could obtain an estimate of a dynamic system at an instant in time. The estimate might be obtained a small time after the measurements because of communication delays but it would be an estimate of the state of the system at the instant the measurements were made. There are several issues that must be addressed. One is the need for redundancy to eliminate bad data and the other is how many PMUs are required. At one extreme if there was a PMU at every bus we would be measuring the state not estimating it. The loss of a measurement in such a case would only mean we lost information about the bus in question but still had knowledge of all other buses.

The first observation is that a PMU in a substation could easily have access to line currents in addition to the bus voltage. Sampling both voltages and currents at the same sampling instants would mean that all phasors would be on the same reference. With a model of the transmission line the knowledge of the line current can be used to compute the voltage at the other end of the line. Measuring line currents can extend the voltage measurements to buses where no PMU is installed. With a large number of PMUs the redundancy issue is addressed. On the other hand, the smallest number of PMUs needed to indirectly measure all the bus voltages and the optimum PMU location to achieve this has been a subject of a number of papers [5–7].

To begin the linear formulation, consider the pi equivalent shown in Figure 7.9 [8–10].

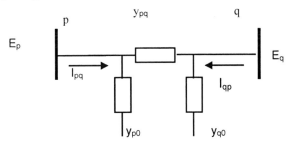

Fig. 7.9 Pi equivalent for a transmission line.

$$\begin{bmatrix} E_p \\ E_q \\ I_{pq} \\ I_{qp} \end{bmatrix} = \begin{bmatrix} 1 & 0 \\ 0 & 1 \\ y_{pq} + y_{p0} & -y_{pq} \\ -y_{pq} & y_{pq} = y_{q0} \end{bmatrix} \begin{bmatrix} E_p \\ E_q \end{bmatrix} \tag{7.35}$$

A current measurement bus incidence matrix is defined in a manner similar to the element bus incidence matrix. It has as many rows as measurements of currents and as many columns as there are buses (excluding ground). Two other matrices are needed as shown in Figure 7.10. If m is the number of current measurements, n the number of lines measured, p the number of buses with voltages measurements, and q the number of buses in the system then A is an $m \times q$ incidence matrix and y is an $m \times m$ diagonal matrix of admittances.

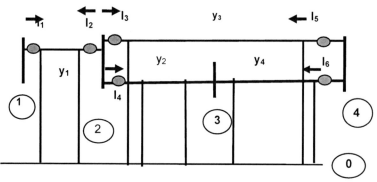

Fig. 7.10 An example with six (m) current measurements on four (n) lines, three (p) voltage measurements and four (q) buses.

The matrices are

$$
A = \begin{bmatrix}
1 & -1 & 0 & 0 \\
-1 & 1 & 0 & 0 \\
0 & 1 & 0 & -1 \\
0 & 1 & -1 & 0 \\
0 & -1 & 0 & 1 \\
0 & 0 & -1 & 1
\end{bmatrix}, \text{ and } \mathbf{y} = \begin{bmatrix}
y_1 & 0 & 0 & 0 & 0 & 0 \\
0 & y_2 & 0 & 0 & 0 & 0 \\
0 & 0 & y_3 & 0 & 0 & 0 \\
0 & 0 & 0 & y_4 & 0 & 0 \\
0 & 0 & 0 & & y_5 & 0 \\
0 & 0 & 0 & & 0 & y_6
\end{bmatrix}, \quad (7.36)
$$

with the shunt branches for each pi section denoted by the subscript 0.

$$
\mathbf{y_s} = \begin{bmatrix}
y_{10} & 0 & 0 & 0 \\
0 & y_{10} & 0 & 0 \\
0 & y_{30} & 0 & 0 \\
0 & y_{20} & 0 & 0 \\
0 & 0 & 0 & y_{30} \\
0 & 0 & 0 & y_{40}
\end{bmatrix} \quad (7.37)
$$

Then the measurement vector composed of p voltages and m currents can be written as

$$
\mathbf{z} = \begin{bmatrix} \mathbf{II} \\ \mathbf{yA + y_s} \end{bmatrix} [\mathbf{E_b}] = \mathbf{BE_B}, \quad (7.38)
$$

where $\mathbf{II}$ is a unit matrix from which rows corresponding to missing bus voltages are removed. For the example in Figure 7.10

$$yA + y_s = \begin{bmatrix} y_1 + y_{10} & -y_1 & 0 & 0 \\ -y_1 & y_1 + y_{10} & 0 & 0 \\ 0 & y_3 + y_{30} & 0 & -y_3 \\ 0 & y_2 + y_{20} & -y_2 & 0 \\ 0 & -y_3 & 0 & y_3 + y_{30} \\ 0 & 0 & -y_4 & y_4 + y_{40} \end{bmatrix} \tag{7.39}$$

or

$$\begin{bmatrix} E_1 \\ E_2 \\ E_4 \\ I_1 \\ I_2 \\ I_3 \\ I_4 \\ I_5 \\ I_6 \end{bmatrix} = \begin{bmatrix} 1 & 0 & 0 & 0 \\ 0 & 1 & 0 & 0 \\ 0 & 0 & 0 & 1 \\ y_1 + y_{10} & -y_1 & 0 & 0 \\ -y_1 & y_1 + y_{10} & 0 & 0 \\ 0 & y_3 + y_{30} & 0 & -y_3 \\ 0 & y_2 + y_{20} & -y_2 & 0 \\ 0 & -y_3 & 0 & y_3 + y_{30} \\ 0 & 0 & -y_4 & y_4 + y_{40} \end{bmatrix} \begin{bmatrix} E_1 \\ E_2 \\ E_3 \\ E_4 \end{bmatrix} . \tag{7.40}$$

The equations are linear and Eq. (7.40) is in the form $z = B\,E_B$:

$$\hat{x} = (B^T W^{-1} B)^{-1} B^T W^{-1} z = Mz . \tag{7.41}$$

Unlike the earlier state estimator, this equation is linear and hence no iterations are needed. As soon as the measurements are obtained, the estimate is obtained by matrix multiplication. The matrix M which converts the measurements to the state estimate is constant as long as the bus structure does not change. It can be computed off-line, and stored for real-time use. Under certain conditions of measurement configuration, the matrix M becomes real, simplifying the computations even further [8–10].

7.5.2 An alternative for including Phasor measurements

An alternate procedure for incorporating the phasor measurements into a conventional estimator is presented in [11]. If the traditional state estimator is in place then rather than the substantial changes required for the method in Section 7.5.1 it is possible to consider the phasor measurements sequentially with the traditional SCADA scan; that is, form the conventional estimate, take it and its covariance matrix and then imagine the phasor measurements as in Section 7.5.1.as an addition. We can imagine combining the two with a measurement equation of the form

$$\begin{bmatrix} E^{(1)} \\ S_2 \end{bmatrix} = \begin{bmatrix} I \\ H_2 \end{bmatrix} [E] \quad \mathrm{Cov} \left(\begin{bmatrix} E^{(1)} \\ S_2 \end{bmatrix} \right) = \begin{bmatrix} H_1^T W_1^{-1} H_1 & 0 \\ 0 & W_2 \end{bmatrix}, \qquad (7.42)$$

where $E^{(1)}$ is the estimate from the conventional estimator and S_2 is the residual from the linear measurements. The old covariance matrix is in polar coordinates while the new must be converted from rectangular to polar. It can be shown that the solution to Eq. (7.42) is the same as the solution of

$$\begin{bmatrix} S_1 \\ S_2 \end{bmatrix} = \begin{bmatrix} H_1 \\ H_2 \end{bmatrix} [E] \quad \mathrm{cov} \left(\begin{bmatrix} S_1 \\ S_2 \end{bmatrix} \right) = \begin{bmatrix} W_1 & 0 \\ 0 & W_2 \end{bmatrix}, \qquad (7.43)$$

which is a nonlinear hybrid estimate handling both the traditional SCADA measurements and the phasor measurements.

7.5.3 Incomplete observability estimators

One of the disadvantages of traditional state estimators is that at the very minimum a complete tree of the network must be monitored in order to obtain a state estimate. The phasor-based estimators have the advantage that each measurement can stand on its own, and a relatively small number of measurements can be used directly if the application requirements could be met. For example, consider the problem of controlling oscillations between two systems separated by great distance. In this case, only two measurements would be sufficient to provide a useful feedback signal.

But in terms of a state estimator application using only PMUs the obvious question is how many PMUs need to be installed in order to measure the state of the system using line currents as discussed in Section 7.5.1. Given the number of lines connecting to each node in a power system is approximately 3 it is clear that a PMU is not necessary at every bus.

The light gray buses in Figure 7.12 are unobservable to a depth of 1 in the sense that they are only one bus away from an observable bus. It is possible to imagine having so few PMUs that depths of unobservability of 2 or 3 or more were achieved. Algorithms to find PMU placements to minimize the number of PMUs for a given depth have been developed [5]. The complete observability case has been approached in a number of

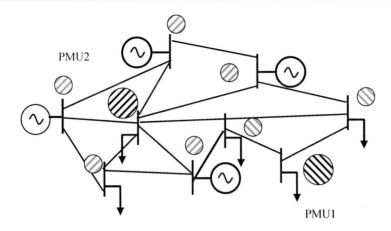

Fig. 7.11 Two PMUs used to observe a nine-bus system. The larger circles represent PMUs while the smaller circles are shaded to indicate which PMU is responsible for the current measurement.

ways with a consensus that PMUs are required at approximately one-third of the buses to obtain complete observabilty. Some examples are given in Table 7.2

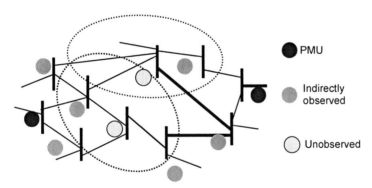

Fig. 7.12 Unobservable buses.

Incidence matrices

The techniques used in [5] involve enumerating trees and can become time consuming as system size increases. The number of possible trees for a 1500-bus system is overwhelming. The idea of searching a tree is appealing when attempting to determine PMU locations which have a certain depth of observability. If only complete observability is of interest the techniques in [6, 7] are more efficient. These calculations involve integer programming and the network incidence matrix. The incidence matrix approach can also be extended to degrees of observability. Consider the network graph shown in Figure 7.13. The incidence matrix is a square matrix with

Table 7.2 Observability results for a few systems [5]

Test system	Size (Buses/Lines)	Complete observability	Depth 1	Depth 2	Depth 3
IEEE 14 Bus	(14,20)	3	2	2	1
IEEE 30 Bus	(30,41)	7	4	3	2
IEEE 57 Bus	(57,80)	11	9	8	7
System α	(270,326)	90	62	56	45
System β	(444,574)	121	97	83	68

the dimension of the number of buses. There is a one on each diagonal and a one in the ij the position if bus i is connected to bus j. The matrix for the network is given by Eq. (7.44). If we imagine placing a PMU at bus 3, for example we would learn the voltages at buses 2, 3, 4, and 6 which happens to be the non-zero entries in column 3 of $\mathbf{A}$.

$$
\mathbf{A} = \begin{bmatrix}
1 & 1 & 0 & 0 & 0 & 1 & 0 & 0 \\
1 & 1 & 1 & 0 & 0 & 0 & 1 & 0 \\
0 & 1 & 1 & 1 & 0 & 1 & 0 & 0 \\
0 & 0 & 1 & 1 & 1 & 0 & 0 & 0 \\
0 & 0 & 0 & 1 & 1 & 0 & 0 & 1 \\
1 & 0 & 1 & 0 & 0 & 1 & 0 & 0 \\
0 & 1 & 0 & 0 & 0 & 0 & 1 & 1 \\
0 & 0 & 0 & 0 & 1 & 0 & 1 & 1
\end{bmatrix}
\tag{7.44}
$$

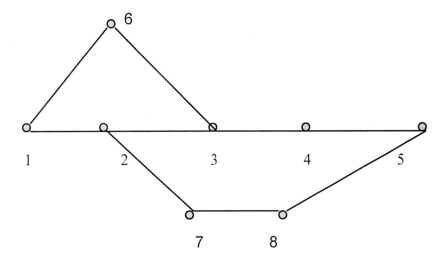

Fig. 7.13 A network graph.

In [6] the problem of locating the minimum number of PMUs to completely observe the network is stated as an integer programming problem
in the form

$$\min \mathbf{f}^T \mathbf{x}$$

$$\text{subject to } \mathbf{Ax} \ge \mathbf{0}, \, x_i = 1 \text{ or } 0.$$ (7.45)

$$\mathbf{f}^T = \begin{bmatrix} 1 & 1 & 1 & \cdots & 1 \end{bmatrix}$$

The form in Eq. (7.45) is the simplest binary integer programming problem. Equality constraints can be added and conventional injection and flow
measurements can be accommodated [7]. The depth of observability calculation, however, was not considered in the approach. It is surprisingly easy
to include by considering the effect of taking powers of the $\mathbf{A}$ matrix. In
[12] it is stated that

Theorem *The ij entry in the nth power of the incidence matrix for any
graph or diagraph is exactly the number of different paths of length n, beginning at vertex i and ending at vertex j.*

The proof is by induction and is easy to see in our example. The signum of
$\mathbf{A}^2$ is the incidence matrix of another graph which has branches added to
the graph in Figure 7.13 as shown in Figure 7.14. The network associated
with $\mathbf{A}^4$ for this example has all nodes connected to all other nodes. In a
"small world" context the example has four degrees of separation, that is,

every node can be reached from any other node by going through four branches.

$$
A^2 = \begin{bmatrix}
3 & 2 & 2 & 0 & 0 & 1 & 1 & 0 \\
2 & 4 & 2 & 1 & 0 & 2 & 2 & 1 \\
2 & 2 & 4 & 2 & 1 & 2 & 1 & 0 \\
0 & 1 & 2 & 3 & 2 & 1 & 0 & 1 \\
0 & 0 & 1 & 2 & 3 & 0 & 1 & 2 \\
2 & 2 & 2 & 1 & 0 & 3 & 0 & 0 \\
1 & 2 & 1 & 0 & 1 & 0 & 3 & 2 \\
0 & 1 & 0 & 1 & 2 & 0 & 2 & 3
\end{bmatrix}.
\tag{7.46}
$$

The 1–5 path is such an example. In terms of depth of unobservability, a single PMU at nodes 2, or 3, or 4, or 7 gives a depth of unobservability of 3. The number of paths is not important in placing PMUs so the sgn(A^n) plays the role of A in the integer programming formulation for the depth of observability problem.

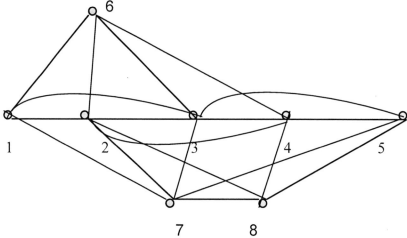

Fig. 7.14 The graph of sgn (A*A).

A common problem is to determine the optimum phased deployment of PMUs. There is usually a limited annual budget to put a certain number of PMUs in place each year. The ultimate goal is to have a completely observable system of measurements at the end of a multiyear time window. It would be best if the PMUs installed in the first year represented a good choice given the small number involved. For example, a set of PMUs that gave some degree of unobservability in each year with a progression to complete observability at the end of period would be ideal. The incidence matrix approach offers a convenient solution to this problem. Suppose $\mathbf{x_0}$ is a solution of the complete observability calculation in Eq. (7.45) and consider

$$\min \mathbf{f}^T \mathbf{x_1} \tag{7.47}$$

$$\text{subject to } [\text{sgn}(\mathbf{AA})]\mathbf{x_1} \geq \mathbf{0},$$

$$(\mathbf{f} - \mathbf{x_0})^T \mathbf{x_1} = 0$$

$$\mathbf{f}^T = \begin{bmatrix} 1 & 1 & 1 & \cdots & 1 \end{bmatrix}$$

The use of sgn($\mathbf{AA}$) assures the solution has depth of observability of 1 and the equality constraint says $\mathbf{x_1}$ must be chosen from the locations in $\mathbf{x_0}$. For a large system a great deal of computation is involved in solving for $\mathbf{x_0}$ while $\mathbf{x_1}$ is much quicker and succeeding solutions are quicker yet. The general form is in Eq. (7.48):

$$\min \mathbf{f}^T \mathbf{x_n} \tag{7.48}$$

$$\text{subject to} [\text{sgn}(\mathbf{A}^{n+1})]\mathbf{x_n} \geq 0$$

$$(\mathbf{f} - \mathbf{x_{n-1}})^T \mathbf{x_n} = 0$$

For the 57-bus system studied in [5] and [6] there are 15 "zero injection buses" (buses with no generation or load). The different treatments of these buses produce slight variation in the results. The technique in [5] produced the numbers in Table 7.2. If the zero injection buses are simple reduced by network equivalencing a 42-bus network is created. The 42-bus network is more "connected" than the original because the elimination of a bus that has connections to both buses p and q produces a new line from p to q. The elimination of 15 buses creates a number of additional lines. The repeated application of Eq. (7.48) produces a nested solution for the reduced network but if the reduced network is too different from the original network there are still unanswered questions for the original network.

A reasonable approach is to limit the network reductions to obvious situations. An example system has 1443 buses and 1929 branches. There are 104 buses that have three or fewer branches connected to them. If they are eliminated by network reduction the number of branches grows to 2178 and the nested solution is shown in Table 7.3. If only the five buses with only two branches are removed the number of branches grows to 1940 and the computation takes considerably longer. Again the results are in Table 7.3.

Table 7.3 Observability results two versions of the 1246-bus system

# Connections to buses removed	(Buses /Lines)	Depth 0	Depth 1	Depth 2	Depth 3	Depth 4	Depth 5
1 or 2 or3	(1339/2178)	433	218	131	74	54	43
1 or 2	(1458/1940)	445	248	135	83	61	42

7.5.4 Partitioned state estimation

Early state estimators were restricted to the highest voltage levels and did not extend to voltages as high as 138 kV in some utilities. The advent of PMU technology made it possible to imagine using the existing conventional estimator supplemented by PMU measurements in portions of the lower voltage system. Again because a few PMU measurements are useful it is not necessary to fully measure the low-voltage network. A second more recent problem of the same type is the seams issue. Two adjoining ISOs with large and elaborate state estimation programs representing immense investment in people and systems would like to combine their state estimates. Typically the people who would actually have to do it have the most reservations. Given the size of an independent system operator (ISO) estimator some reservations are justified. Combining two 30 000 bus estimators which probably work differently and their own quirks is a daunting prospect. Both problems involve the concept of joining state estimators.

Some have suggested that the only optimum answer to these problems is to start over and pool the models and data and buy more computers. If a simpler solution which saves money, time, and frustration can be found it certainly seems worth considering. Consider the system in Figure 7.15. The boundary buses in Figure 7.15 could be the high sides of transformers connecting high- and low-voltage subsystems or the buses at the end of the tie lines between ISOs. An attractive approach to the problem is to let each system use their existing state estimator but recognize that the two estimators do have different references. By including the boundary buses in both systems we can estimate the difference between the two references. If NB is the number of boundary buses, and sub 1 and 2 denote estimates from each side then ϕ is the estimated difference between the two references.

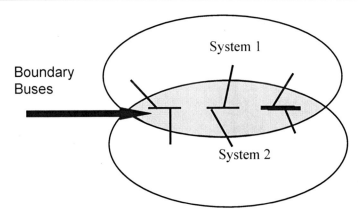

Fig. 7.15 Two systems with boundary buses.

$$\varphi = \frac{1}{NB} \sum_{i=1}^{NB} (\hat{\theta}_{1i} - \hat{\theta}_{2,i}).$$

(7.49)

Example 7.5

The 30 bus system is divided into two subsystems as shown in Figure 7.16. Two estimators are formed using the tie-line flows as sources or

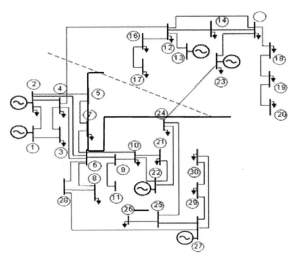

.**Fig. 7.16** The 30-bus system divided into two subsystems.

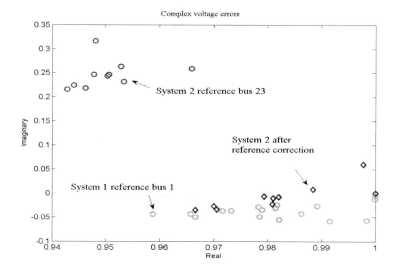

Fig. 7.17 The voltage errors before and after correction.

loads. The reference for system 1 is bus 1 as before but the reference for system 2 is bus 23. The voltage errors are shown in Figure 7.17. The diamonds show the corrected errors in system 2 after the reference is estimated.

7.5.4.1 Nonlinear *residual minimization method*

A study of alternative approaches to integrating adjoining estimators is in [13] where the solution that combines small computational burden, accuracy, and robustness was chosen as the "nonlinear residual minimization" (NLRM). Let the subscript C denote calculated quantities given and N be the total number of tie lines between the two systems. The calculated quantities for a line from node a to node b are given by Eqs. (7.50) and (7.51). In the method (NLRM) the angle difference between systems, ϕ, was chosen to minimize $J(\phi)$ in Eq. (7.52). In Eq. (7.52) the subscript m denotes measured quantities.

$$P_C = \frac{\left|V_a\right|^2}{Z_{ab}}\cos(\theta_z) - \frac{\left|V_a\right|\left|V_b\right|}{Z_{ab}}\cos(\theta_z + \delta_{ab} - \phi) \quad (7.50)$$

$$Q_C = \frac{|V_a|^2}{Z_{ab}}\sin(\theta_z) + \frac{|V_a|^2 B_{ab}}{2} - \frac{|V_a||V_b|}{Z_{ab}}\sin(\theta_z + \delta_{ab} - \phi) \qquad (7.51)$$

$$J(\varphi) = \left\| \begin{array}{c} \dfrac{P_{m1} - P_{C1}}{\sigma_{P_1}} \\[2ex] \dfrac{Q_{m1} - Q_{C1}}{\sigma_{Q\,1}} \\[2ex] \dfrac{P_{m2} - P_{C2}}{\sigma_{P_2}} \\[2ex] \dfrac{Q_{m2} - Q_{C2}}{\sigma_{Q\,2}} \\[1ex] \vdots \\[1ex] \dfrac{P_{mN} - P_{CN}}{\sigma_{P_j}} \\[2ex] \dfrac{Q_{mN} - Q_{CN}}{\sigma_{Q\,N}} \end{array} \right\|^2 \qquad (7.52)$$

Using Eq. (7.52) with $\eta = \theta_z + \delta_{ab}$

$$\begin{aligned} \cos(\eta - \phi) &= \cos\eta\cos\phi + \sin\eta\sin\phi \\ \sin(\eta - \phi) &= \sin\eta\cos\phi - \cos\eta\sin\phi \end{aligned} \qquad (7.53)$$

Then the objective function is given by Eq. (7.54):

$$J(\phi) = \|X + Y\cos\phi + Z\sin\phi\|^2 \qquad (7.54)$$

Since X, Y, and Z are constants Eq. (7.54) is equivalent to Eq. (7.55) where A, B, C, D, and E are constants:

$$J(\phi) = A\cos\phi + B\sin\phi + C\cos\phi\sin\phi + D\cos^2\phi + E\sin^2\phi \qquad (7.55)$$

The minimization of Eq. (7.55) involves taking the first derivative with respect to ϕ, equating it to zero and using Newton's method to solve the resulting scalar nonlinear equation.

7.6 Calibration

Calibration as an issue in state estimation predates phasor measurements and is still an issue with phasor measurements. In the previous sections we have modeled the measurement error as a simple additive error. Our numerical examples have used rather small standard deviations expressed as a percent of the measured quantity; that is, if we measured a flow of 10 MW we added a random error with a standard deviation of say 1% of the 10 MW to both the real and imaginary measurements. It is assumed in our previous development that the error has zero mean and if the estimation process is repeated every few minutes the error as a particular measurement will be independent from measurement to measurement and have the same statistical description. Two measurements of a complex flow are as shown in Figure 7.18a. It assumed that one measurement is approximately half the other in magnitude with a slightly different angle. A small circle of error about the correct value with a radius proportional to the size of the measurement is drawn to represent the error. It is a circle in the complex plane because we are assuming the same size error is added to the real and imaginary measurements. In Figure 7.18 the solid lines are the true values and the dashed lines are the hypothetical measurements.

The problem lies with the transducers, the current and potential transformers. They contribute a systematic error as shown in Figure 7.18b. If the true current is I the measured value is γI, where γ is always the same complex number – complex because there is both a magnitude and a phase shift error. While γ is not known when a single estimate is formed it has been suggested that over time these unknown constants could be learned, that is, that the measurements could be calibrated. The circles in Figure 7.18 were large enough for early analog measurements that many solved the problem by just increasing the standard deviation to take the systematic errors into consideration.

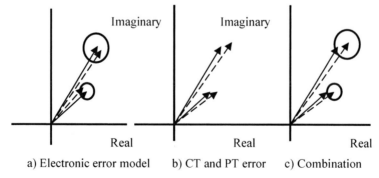

a) Electronic error model   b) CT and PT error   c) Combination

Fig.7.18 Error models for the calibration problem.

Mathematically the model has changed from Eq. (7.22) to

$$z_i = \gamma_i h_i(V, \theta) + \varepsilon \tag{7.56}$$

and the problem is to estimate the γs. Since every CT and PT could contribute a γ it is clear that in the worst the γs cannot be estimated in a single estimate. There could be more unknowns than measurements. The fact that they are constant must be exploited to calibrate the measurements over time. In [14] it is assumed that only some measurements need to be calibrated and the normalized residuals are used to select those. More complicated models than Eq. (7.56) are also required. Calibration of a power flow, for example, would require a quadratic model since both a CT and a PT are involved. If the number of measurements to be calibrated is small enough the unknown γs can be added to the state, the Jacobian adjusted accordingly, and the procedure of section 7.3 followed with an augmented state. However, most calibration techniques involve the use of multiple scans and assume the systematic errors are constant while the data is collected. In [15] it is reported that data was used over a period of several days to perform a calibration.

Another approach is to calibrate CTs and PTs separately for different system conditions [16]. For a lightly loaded system currents are small and systematic CT errors make little contribution. If some voltages measurements are known to be good (tie-line voltages, for example) then unbiased current measurements and a few unbiased voltage measurements are sufficient to estimate the system state and the γ_i for the remaining voltages. A linear estimator using only PMU measurements is considered in [16]. One approach is to augment state to include the complex bus voltages and the γ_i for the uncalibrated voltage measurements. If there are Nb buses and Nl lines then a complete set of linear measurements would include Nb complex voltages and 2Nl complex current measurements. If Nt of the voltages are already calibrated then there are (Nb – Nt) γs and Nb complex voltages to estimate with Nb + 2Nl measurements. The problem is nonlinear because γ_i appears as a multiplier. Using the system of Section 7.5.1 as an example if we assume the voltage at bus one is free of systematic error but E_2 and E_4 have systematic errors then Eq. (7.40) can be written as

$$
\begin{bmatrix} E_1 \\ E_2 \\ E_4 \\ I_1 \\ I_2 \\ I_3 \\ I_4 \\ I_5 \\ I_6 \end{bmatrix} = \begin{bmatrix} 1 & 0 & 0 & 0 \\ 0 & \gamma_2 & 0 & 0 \\ 0 & 0 & 0 & \gamma_4 \\ y_1 + y_{10} & -y_1 & 0 & 0 \\ -y_1 & y_1 + y_{10} & 0 & 0 \\ 0 & y_3 + y_{30} & 0 & -y_3 \\ 0 & y_2 + y_{20} & -y_2 & 0 \\ 0 & -y_3 & 0 & y_3 + y_{30} \\ 0 & 0 & -y_4 & y_4 + y_{40} \end{bmatrix} \cdot \begin{bmatrix} E_1 \\ E_2 \\ E_3 \\ E_4 \end{bmatrix}.
\tag{7.58}
$$

If we take the line admittances to be equal and the shunt admittances as 10% of the line admittances we get a B matrix as in the given matlab m file CWL.m. The nonlinear estimation problem converges in two iterations with the matrix M in the form of Eq. (7.58). The number β and δ in Eq. (7.58) express a dependence of γ_2 and γ_4 on the residuals in E_2 and E_4. The estimation of voltage γ_i under light load does not require accumulation of many data points. One light-load measurement is sufficient to calibrate the voltage measurements given some number of voltage are already calibrated. Clearly more measurements at the lightly loaded case will improve the estimate of the voltage multipliers, where n_1 and n_2 were learned from the light-load case and are assumed known. The measurement vector $\mathbf{z}$ is the same measurement as in the lightly loaded case.

Given the voltage multipliers have been estimated the heavy-load case can be examined to estimate the CT γ_is. The issue that must be addressed is that the number of measurements is not really adequate to estimate the CT

$$
M = \begin{bmatrix} x & \beta & 0 & x & x & x & x & x & x \\ x & 0 & \delta & x & x & x & x & x & x \\ x & 0 & 0 & x & x & x & x & x & x \\ x & 0 & 0 & x & x & x & x & x & x \\ x & 0 & 0 & x & x & x & x & x & x \\ x & 0 & 0 & x & x & x & x & x & x \\ x & 0 & 0 & x & x & x & x & x & x \end{bmatrix}
\tag{7.59}
$$

multipliers with one measurement. In the example if γ_2 and γ_4 are assumed known but γs are introduced for the current measurements then Eq. (7.57) becomes

$$
\mathbf{z} = \begin{bmatrix}
1 & 0 & 0 & 0 \\
0 & n_2 & 0 & 0 \\
0 & 0 & 0 & n_4 \\
\gamma_1(y_1 + y_{10}) & -\gamma_1(y_1) & 0 & 0 \\
-\gamma_2(y_1) & \gamma_2(y_1 + y_{10}) & 0 & 0 \\
0 & \gamma_3(y_3 + y_{30}) & 0 & -\gamma_3(y_3) \\
0 & \gamma_4(y_2 + y_{20}) & -\gamma_4(y_2) & 0 \\
0 & -\gamma_5(y_3) & 0 & \gamma_5(y_3 + y_{30}) \\
0 & 0 & -\gamma_6(y_4) & \gamma6(y_4 + y_{40})
\end{bmatrix} \mathbf{E} \qquad (7.59)
$$

loaded case (three voltages and six currents) but now there are 10 states to be estimated: the six multipliers and the four voltages. If a second set of measurements are taken later and combined with the first nine measurements then we would have 18 measurements and 14 unknowns (four voltages from the first measurement, four more voltages from the second, and six unknown γs).

The Jacobian for two measurements is given in Eq. (7.60), where

$$
\mathbf{H} = \begin{bmatrix}
\mathbf{N} & \mathbf{0} & \mathbf{0} \\
\mathbf{\Gamma Y} & \mathbf{0} & \mathbf{diag(YE^{(1)})} \\
\mathbf{0} & \mathbf{N} & \mathbf{0} \\
\mathbf{0} & \mathbf{\Gamma Y} & \mathbf{diag(YE^{(2)})}
\end{bmatrix} \quad for \quad \mathbf{x} = \begin{bmatrix}
\mathbf{E}^{(1)} \\
\mathbf{E}^{(2)} \\
\gamma
\end{bmatrix} . \qquad (7.60)
$$

Γ is a diagonal matrix of the estimated current γs, $\mathbf{Y}$ is the matrix Eq. (7.42), $\mathbf{E}^{(1)}$ and $\mathbf{E}^{(2)}$ are voltages estimates at the two measurement, and N is the diagonal matrix in Eq. (7.61).

$$
\mathbf{N} = \begin{bmatrix}
1 & 0 & 0 \\
0 & n_1 & 0 \\
0 & 0 & n_2
\end{bmatrix} \qquad (7.61)
$$

MATLAB File CWL.m

```
%CWL.m    Calibration with load 7-6-07
E=[1 1 1 1]';
eg=0.1*randn(2,1)
%The gammas are random but chosen from a population
with sigms=10%
g20=1+eg(1);
g40=1+eg(2);
yt=[g20 g40 E']'; % the true state
ee=0.03*randn(9,1); %3 measurement errors on voltages
and currents

BB=[1 0 0 0
    0 g20 0 0
    0 0 0 g40
    1.1 -1 0 0
    -1 1.1 0 0
    0 1.1 0 -1
    0 1.2 -1 0
    0 -1 0 1.1
    0 0 -1 1.1];
z=BB*E+ee; % measurements

g2=1;
g4=1;
% 3 iterations
for i=1:3
B=[1 0 0 0
    0 g2 0 0
    0 0 0 g4
    1.1 -1 0 0
    -1 1.1 0 0
    0 1.1 0 -1
    0 1.2 -1 0
    0 -1 0 1.1
    0 0 -1 1.1];


H=0.0*ones(9,6);
H(1:9,3:6)=B;
H(1:3,1:2)=[0 0
    E(2) 0
    0 E(4)];
M=(inv(H'*H))*H'
yn=[g2 g4 E']'+M*(z-B*E);
```

```
 g2=yn(1);
g4=yn(2);
E=yn(3:6,1);

eee(i)=sqrt((yt-yn)'*(yt-yn));
end
plot(eee)
```

7.7 Dynamic estimators

Before the introduction of phasor measurements the idea of tracking the state of the power system was introduced [17]. It was assumed that state of the system could be modeled as Eq. (7.62), where $x(k)$ is the state at the kth time step, Δt is the time step, r is a maximum rate of change vector, $z(k)$ is the measurement, and $v(k)$ is the measurement error.

$$x(k+1) = x(k) + (\Delta t)r$$
$$z(k) = Hx(k) + v(k) \tag{7.62}$$

If the term $(\Delta t)r$ is denoted $w(k)$, and $w(k)$ and $v(k)$ are modeled as zero mean, independent, Gaussian processes then the problem is essentially a Kalman filtering problem [18]. The model incorporates a random evolution of the state through $w(k)$ but loses power system details in the form of the measurements and the linear model. Phasor measurements with time-tags could produce a modern equivalent of the tracking estimator in [17]. PMU measurements could be integrated with the existing data scans. It is important to model the system dynamics that are being tracked. There are certainly situations in which the system responds in a roughly predictable manner, the morning load pick-up, for example, is quite dependable as are other daily events. An estimator that assumed that the changes in load were linear in time with unknown rates of change (which could themselves be estimated) would seem to be a considerable improvement over the static assumption. This is, in effect, a matter of placing the phasor measurements correctly in time within the SCADA data window. The problem of skew of PMU measurement integrated with conventional data is difficult. If the data scan for conventional measurements takes T seconds and the PMU data can be located anywhere in the T second interval, should the phasor measurements be distributed in the window or concentrated at one point? If at one point, should that point be at the beginning, the middle, or the end of the observation window?

The preceding could ultimately lead to an estimator based on more PMU data that could follow the state through transient swings. With PMU measurements available as often as every two cycle, transient swings with one second periods could be tracked. There would be some delay but a true dynamic estimate that was a few hundred millisecond delayed is imaginable.

It is also possible to consider the use of real-time phasor measurements in the estimation of parameters or validation of models of components [8]. The use of phasors and local frequency in estimating machine parameters and internal states can also be approached as a Kalman filtering problem. [19] At the level of validating system models, one could replace parts of the system with observed phasor voltage and currents as time-dependent sources, and then improve the model of other parts of the system.

References

1. Allemong, J.J., et. al., "A fast and reliable state estimation algorithm for AEP's new control center", IEEE Transactions on PAS, Vol. 101, No. 4, April 1982, pp 933–944.
2. Dopazo, J.F., et. al., "Implementation of the AEP real-time monitoring system", IEEE Transactions on PAS, Vol. 95, No. 5, September/October 1975, pp 1618–1529.
3. Handshin, E., et.al., "Bad data analysis for power system static state estimation", IEEE Transactions on PAS, Vol. 94, No. 2, March/April 1975, pp 329–337.
4. Monticelli, A., Wu, F.F., and Yen, M., "Multiple bad data identification for state estimation using combinatorial optimization", IEEE PAS-90, November/December 1971, pp 2718–2725.
5. Nuki, R.F. and Phadke, A.G., "Phasor measurement placement techniques for complete and incomplete observability", IEEE Transactions on Power Delivery, Vol. 20, No. 4, October 2005, pp 2381–2388.
6. Gou, B. and Abur, A., "An improved measurement placement algorithm for network observability", IEEE Transactions on Power Systems, Vol. 16, No. 4, November 2001, pp 819–824.
7. Abur, A. "Optimal placement of phasor measurements units for state estimation", PSERC Publication 06-58, October 2005.
8. Phadke, A.G. and Thorp, J.S., "Computer Relaying for Power Systems", Research Studies Press, Somerset, England, 1988.
9. Thorp, J.S., Phadke, A.G., and Karimi, K.J., "Real-time voltage phasor measurements for static state estimation", IEEE Transactions on PAS, Vol. 104, No. 11, November 1985, pp 3098–3107.

10. Phadke, A.G., Thorp, J.S., and Karimi, K.J., "State estimation with phasor measurements", IEEE Transactions on PWRS, Vol. 1, No. 1, February 1986, pp 233–241.
11. Zhou, M., et al., "An alternative for including phasor measurements in state estimation", IEEE Transactions on Power Systems, Vol. 21, No. 4, November 2006, pp 1930–1937.
12. Korevaar, N. "Incidence is no coincidence", University of Utah Math Circle, October 2002.
13. Jeffers, R., "Wide area state estimation techniques using phasor measurement data" Virginia Tech Report prepared for Tennessee Valley Authority, March 2007.
14. Zhong, S. and Abur, A., "Combined state estination and measurement calibration", IEEE Transactions on Power Systems, Vol. 20, No. 1, February 2005, pp 458–465.
15. Adibi, M.M. and Kafka, R.J., "Minimization of uncertainties in analog measurements for use in state estimation", IEEE Transactions on Power Systems, Vol. 5, No. 3, August 1990, pp 902–910.
16. Zhou, M., Ph.D. Dissertation, Virginia Tech March 2008, "Phasor Measurement Unit Calibration and Applications in State Estimation".
17. Debs. A.S., Larson, R.E., "A dynamic estimator for tracking the state of a power system", IEEE Transactions on Power Apparatus Systems, Vol. PAS-893, No. 7, September/October 1970, pp 1670–1678.
18. Gelb, A., "Applied Optimal Filtering", MIT Press, Cambridge, 1974.
19. Pilay, P., Phadke, A.G., Linders, D.K., Thorp, J.S., "State Estimation for a Synchronous Machine: Observer and Kalman Filter Approach", Princeton Conference, 1987.

Chapter 8 Control with Phasor Feedback

8.1 Introduction

Prior to the introduction of real-time phasor measurements power system control was essentially by local signals. Feedback control with such locally available measurements is widely used in controlling machines. In other situations, control action was taken on the basis of a mathematical model of the system without actual measurement of the system. The advent of phasor measurements allows the consideration of control based on the measured value of remote quantities. It is expected that such control will be less dependent on the model of the system being controlled. The fact that most such phenomena are relatively slow is an encouraging factor for deploying phasor measurement units (PMUs). The latency of the phasor measurement process is not important when the process frequencies are in the 0.2–2.0 Hz range. The phasor data would be time-tagged so that control would be based on the actual state of the system a short time earlier. The frequencies are representative of the electromechanical oscillations, transient stability, and certain overload phenomena. The frequency of measurements is expected to be of the order of 15–30 Hz, which is certainly sufficient to handle the control task.

Studies of control of high voltage direct current (HVDC) systems, excitation control, power system stabilizers, and flexible alternating current transmission systems (FACTS) control will be described in the next few sections. All of these applications share common features. The actual processes are inherently nonlinear because they involve real power and there are never enough measurements to totally describe the dynamical system in the same detail as a typical aerospace application. The next section provides a framework to investigate these problems.

8.2 Linear optimal control

A general control design used in this section is shown in Figure 8.1.

A.G. Phadke, J.S. Thorp, *Synchronized Phasor Measurements and Their Applications*, DOI: 10.1007/978-0-387-76537-2_8, © Springer Science+Business Media, LLC 2008

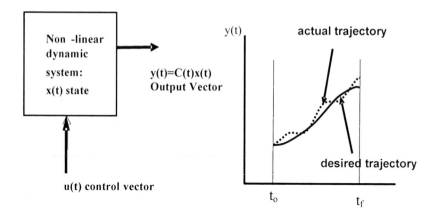

Fig. 8.1 Controller design.

The control law will minimize the difference between the output desired trajectory $\mathbf{y}(t)$ and the desired trajectory $\mathbf{y}_d(t)$. The problem is well studied for linear systems [1]. For the system in Eq. (8.1)

$$\dot{\mathbf{x}}(t) = \mathbf{A}(t)\mathbf{x}(t) + \mathbf{B}(t)\mathbf{u}(t)$$
$$\mathbf{y}(t) = \mathbf{C}(t)\mathbf{x}(t).$$

(8.1)

The finite interval problem is to minimize J by choice of $\mathbf{u}(t)$ where J is in Eq. (8.2):

$$J = \frac{1}{2}\int_{t_0}^{t_f}\left\{[\mathbf{y}(t) - \mathbf{y}_d(t)]^T \mathbf{Q}[\mathbf{y}(t) - \mathbf{y}_d(t)] + \mathbf{u}^T(t)\mathbf{R}\,\mathbf{u}(t)\right\} dt +$$
$$\frac{1}{2}[\mathbf{y}(t_f) - \mathbf{y}_d(t_f)]^T \mathbf{H}[\mathbf{y}(t_f) - \mathbf{y}_d(t_f)]$$

(8.2)

The performance index has weighting throughout the interval on the error in achieving the desired output trajectory. The error is weighted by the matrix Q and the control effort required is weighted by the matrix R. Typically Q and R are diagonal with weights that take different units and maximum values of the variables into consideration. For example, if the states included the altitude of an aircraft in feet and the angle of attack in radians, then very different diagonal weights are required to keep the feet variables from swamping out the radian variable terms. The solution is given by Eq. (8.3) where the feedback gain matrix $K(t)$ and forcing term $g(t)$ are given in Eq. (8.4).

$$\mathbf{u}^*(t) = \mathbf{R}^{-1}\mathbf{B}^T\left[g(t) - \mathbf{K}(t)\mathbf{x}(t)\right]\!] \tag{8.3}$$

$$\dot{\mathbf{K}}(t) = -\mathbf{A}^T\mathbf{K} - \mathbf{K}\mathbf{A} + \mathbf{K}\mathbf{B}\mathbf{R}^{-1}\mathbf{B}^T\mathbf{K} - \mathbf{C}^T\mathbf{Q}\mathbf{C}$$
$$\text{with} \quad \mathbf{K}(t_f) = \mathbf{C}^T(t_f)\mathbf{H}\mathbf{C}(t_f)$$
$$\dot{\mathbf{g}}(t) = -[\mathbf{A}^T - \mathbf{K}\mathbf{B}\mathbf{R}^{-1}\mathbf{B}^T]\mathbf{g}(t) - \mathbf{C}^T\mathbf{Q}\mathbf{y}_d$$
$$\text{with} \quad \mathbf{g}(t_f) = \mathbf{C}^T(t_f)\mathbf{H}\mathbf{y}_d(t_f) \tag{8.4}$$

The matrix $\mathbf{K}$ in Eq. (8.4) satisfies the so-called Riccati equation [1]. It involves considerable computation and is frequently approximated by its steady-state solution which can be obtained with only algebraic manipulations. The steady state can be reached if the time interval $t_f - t_0$ is long enough. Note that the control $\mathbf{u}^*$ applied to the original state equation results in a closed loop system:

$$\dot{\mathbf{x}}(t) = [\mathbf{A}(t) - \mathbf{B}(t)\mathbf{R}^{-1}\mathbf{B}^T\mathbf{K}(t)]\mathbf{x}(t) + \mathbf{B}(t)\mathbf{R}^{-1}\mathbf{B}^T\mathbf{g}(t),$$
$$\mathbf{y}(t) = \mathbf{C}(t)\mathbf{x}(t). \tag{8.5}$$

The A matrix from Eq. (8.1) has now been transformed to the closed loop plant with "A" matrix as $\mathbf{A}\text{-}\mathbf{B}\mathbf{R}^{-1}\mathbf{B}^T\,\mathbf{K}$. The transpose of that "A" matrix appears in the differential equation for $\mathbf{g}(t)$. So the $\mathbf{g}(t)$ equation in Eq. (8.4) is the adjoint of the closed loop system. It is also important to recognize that both equations in Eq. (8.4) are solved backward in time, that is, the terminal conditions are given rather than initial conditions. These points are important to understand the nature of the approximation as discussed in the next section.

8.3 Linear optimal control applied to the nonlinear problem

Given phasor measurements of the system, it is possible to measure the difference between the states of system we are actually controlling and the state of a model. If the actual system is given by Eq. (8.6) then we can have a simpler model in mind (a linearized or reduced order model):

$$\dot{\mathbf{x}}(t) = \mathbf{F}(\mathbf{x}(t), \mathbf{u}(t), t),$$
$$\mathbf{y}(t) = \mathbf{C}(t)\mathbf{x}(t). \tag{8.6}$$

$$\dot{\mathbf{x}}(t) = \mathbf{A}(t)\mathbf{x}(t) + \mathbf{B}(t)\mathbf{u}(t) + \mathbf{f}(t),$$
$$\text{where} \quad \mathbf{f}(t) = \mathbf{F}(\mathbf{x}(t), \mathbf{u}(t), t) - \mathbf{A}(t)\mathbf{x}(t) \tag{8.7}$$

The term $\mathbf{f}(t)$ in Eq. (8.7) is the difference between the derivative of the state in the actual system and in the linear model. Assuming $\mathbf{f}(t)$ is known for the time being, the solution to the optimal control problem is given by the equation in Eqs. (8.3) and (8.4) with one small addition. The differential equation for $\mathbf{g}$ has an additional term at the end depending on $\mathbf{f}$ as given in Eq. (8.8).

$$\dot{\mathbf{g}}(t) = -[\mathbf{A}^T - \mathbf{KBR}^{-1}\mathbf{B}^T]\mathbf{g}(t) - \mathbf{C}^T\mathbf{Q}\mathbf{y}_d + \mathbf{Kf},$$
$$\text{with} \quad \mathbf{g}(t_f) = \mathbf{C}^T(t_f)\mathbf{H}\mathbf{y}_d(t_f) \tag{8.8}$$

It is convenient that the Riccati equation is not affected so it can be computed off line and stored. Only Eq. (8.8) must be solved in order to determine the control. The problem is that Eq. (8.8) is solved backward in time and we can compute $\mathbf{f}(t)$ forward in time as shown in Figure 8.2. The solution is to predict $\mathbf{f}(t)$ based on the data that has been stored since the

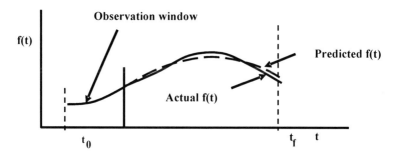

Fig. 8.2 Predicting $\mathbf{f}(t)$.

optimization began. The prediction is worst at the beginning of the interval and improves as time increases. The next few sections give some application of this approximate control design to some systems.

Example 8.1 HVDC system.

In [2] the predictive control was applied to a HVDC system shown in Figure 8.3. The modeling involved combining the network with the HVDC model, generator models, and the exciter models to write the state equations. It was found that the steady-state value of the Riccati equation was acceptable and greatly reduced the computational burden. Even a steady-state value of $g(t)$ was used by finding the steady-state solution of Eq. (8.8) given by Eq. (8.9) (note this value g gives $\dot{g} = 0.$)

$$g = [A^T - KBR^{-1}B^T]^{-1}Kf \qquad (8.9)$$

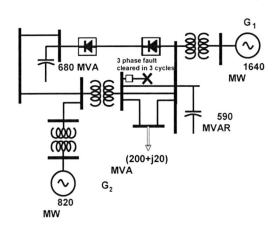

Fig. 8.3 The HVDC system.

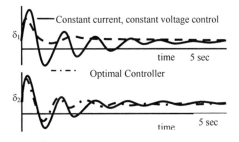

Fig. 8.4 Performance of the HVDC controller.

The prediction of $\mathbf{f}(t)$ was performed with a straight line, obtained by computing the moving average of the previous values of $\mathbf{f}(t)$. The system is shown in Figure 8.3. It has two generators, three transformers with two of them having the off-nominal turns ratios, a two-terminal HVDC link, a load bus, and an infinite bus. A three-phase fault is applied as shown for three cycles and the line removed to clear the fault. The performance of the controller is shown in Figure 8.4. The solid curves are the two rotor angles with a constant current and constant voltage on the HVDC. The dashed curves are the rotor angles obtained with the steady-state Riccati equation and the steady-state $\mathbf{g}$ equation (Eq. 8.9).

Example 8.2 Excitation control.

A centralized excitation controller can be designed using the same technique. If all the real-time phasor data were brought to a central location as in [3] and all the control signals for the generator excitation and governor systems computed using Eq. (8.3) then a centralized excitation controller could be designed (Figure 8.5).

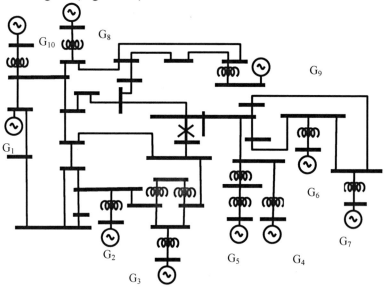

Fig. 8.5 A 10-machine system used for a centralized excitation controller.

The controller state variables correspond to the incremental changes from pre-fault values. The IEEE Type 1 exciter model with an applied auxiliary input signal was used [4]. The desired trajectory corresponds to a desired post-fault equilibrium and the known state of the power system immediately after the fault. Four machine angles from the 39-bus New England system are shown in Figure 8.6 with the angle of machine 10 as the reference with and without feedback control. Again as in the HVDC example steady-state values are used for the Riccati equation and the equilibrium **g** equation.

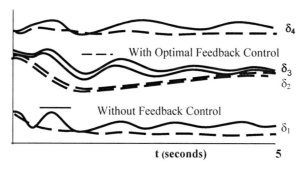

Fig. 8.6 Performance of the excitation control.

Example 8.3 FACTS controller.

A three-machine system with a thyristor-controlled series capacitor (TCSC) is shown in Figure 8.7 [5]. The incremental linear system was obtained by linearizing around the initial operating point. IEEE Type I exciters were used for the machines. Sensitivity analysis was used to locate the TCSC for maximum effect on eigenvalue location. The reactive loading levels were increased until there was a pair of unstable modes as shown in Table 8.1.

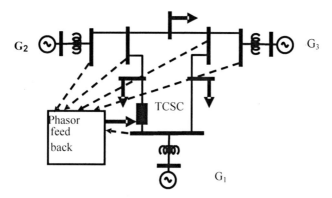

Fig. 8.7 FACTS example system.

The phasor measurements control the compensating reactance of the transmission line with TCSC as shown in Figure 8.8. A PID controller with an input of the angle difference $\delta_1 - \delta_2$ produced an output signal to control the reactance as shown in Figure 8.8.

Table 8.1 Eigenvalues for the system in Figure 8.7

−60.22	−2.5341
−26.5	$0.0021 \pm j0.3043$
−16.57	−1.5769
−9.21	−0.6859
$−5.52454 \pm j3.8478$	$−0.38 \pm j0.0494$
−7.3475	−1.3394
$−2.7664 \pm j0.9461$	−0.1747
	−2.00

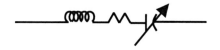

Fig. 8.8 Thyristor-controlled series capacitor (TCSC) control.

The performance of the optimal control solution is shown in Figure 8.9 for a step change in one generator power.

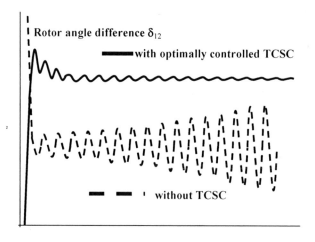

Fig. 8.9 FACTS controller performance.

Example 8.4 Power system stabilizer.

Three control schemes were tested on the four-machine system in Figure 8.10: "automatic voltage regulator" on all four machines, a conventional "power system stabilizer" on one machine as shown, and phasor feedback on the same machine. Modal analysis of the system shows that at tie flow of 158 MW there is an unstable interarea mode. The modes are shown on Table 8.2. The generators were modeled following the two-axis method [6, 4] and detailed models used for the governor, turbine, and constant gain excitation systems.

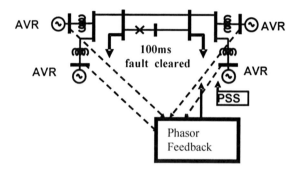

Fig. 8.10 Example system for AVR PSS and phasor feedback comparison.

Table 8.2 Eigenvalues for the example with 158-MW tie flow

Mode	Frequency	Damping ratio	Mode type
$-.05977 \pm j7.0365$	1.1199	0.0849	Local Area 1
$-0.6060 \pm j7.247$	1.1534	0.0833	Local Area 2
$0.0296 \pm j4.1784$	0.665	-0.0071	Inter-area

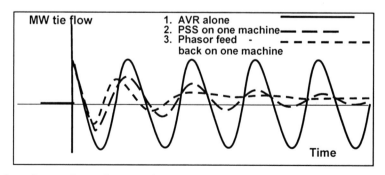

Fig. 8.11 Comparison of automatic voltage regulator (AVR)s, a single PSS, and a phasor-based PSS.

The interarea mode is stable for tie flow of 50 MW but becomes unstable for larger flows between the two areas. A comparison of the three control schemes is given in Figure 8.11. In a separate study the amount of latency that could be tolerated in the phasor measurements for the power system stabilizer was determined to be as much as 150 ms depending on the frequency of the oscillation.

8.4 Coordinated control of oscillations

The power system stabilizers of the preceding example are typically used to control interarea oscillations. These oscillations are low-frequencysmall signal oscillations that seem to be growing in number. A single 0.7-Hz oscillation in the western electric coordinating council (WECC) has been replaced by as many as five frequencies with some as low as 0.2 Hz. Stabilizers are tuned to damp a specific mode and when installed are effective. The difficulty is that as the system changes, the stabilizer is not quite as effective. It is also conjectured that the stabilizers interact with each other to produce new modes. Given the evolving nature of the frequencies and occurrences of the modes it would be best if some strategy could be devised to provide damping for all modes rather than designing specific controllers aimed precisely at presumed modes. Existing approaches have been shown to lack robustness. A parallel to this problem exists in structures both tall buildings and large space structures. In both cases it is desirable to damp vibrations without knowing precisely what form the vibrations will take. Earthquakes and unusual winds for tall buildings and unpredicted disturbances on the space station are examples.

A common solution to the structural engineering problem is the use of the so-called 'collocated control' [7]. It seems that phasor measurements can provide a similar solution to low-frequency interarea oscillations in power systems. The basic idea in the structures problem is to formulate the problem in modal form as in Eq. (8.10):

$$\ddot{\eta} + D\dot{\eta} + \Lambda^2\eta = Bu; \quad u = -Fy; \quad y = B^T\eta, \tag{8.10}$$

where η is the vector of modal coordinates, u is the vector of control inputs, and y is the vector of measurements. The matrix Λ is a diagonal matrix $\Lambda = \mathrm{diag}\{\omega_1^2 \cdots \omega_n^2\}$ with $\omega_1 < \omega_2 < \cdots \omega_n$. We assume the damping is proportional to frequency, $D = 2\alpha\Lambda$, where $\alpha = \mathrm{diag}\{\alpha_1 \cdots \alpha_n\}$ with $\alpha_i < 1$. F is a non-negative definite matrix to be

determined. The open loop eigenvalues of the system
are $-\alpha_i\omega_i \pm j\omega_i\sqrt{1-\alpha_i^2}$. We assume the first K modes are the critical
low-frequency modes for which we wish to provide additional damping.
The term collocated refers to the matrix **B** appearing with both **u** and **y** in
Eq. (8.10). It produces a convenient form for the eigenvalues of the closed
loop system of Eq. (8.10) given in Eq. (8.11):

$$\ddot{\boldsymbol{\eta}} + \mathbf{D(F)}\dot{\boldsymbol{\eta}} + \boldsymbol{\Lambda}^2\boldsymbol{\eta} = 0; \quad \mathbf{D(F)} = 2\alpha\boldsymbol{\Lambda} + \mathbf{BFB}^{\mathrm{T}}. \tag{8.11}$$

The collocated form guarantees that the damping added by the feedback
does no harm even if the system model changes. The term $\mathbf{BFB}^{\mathrm{T}}$ is non-
negative definite and behaves like a multidimensional resistive network.
    An optimal **F** can be considered by mapping the complex plane in which
the eigenvalues of Eq. (8.11) reside. If $\bar{\lambda}$ is a mapping from λ, the eigen
values of Eq. (8.11), given by

$$\bar{\lambda} = \frac{r - z_0 + \lambda}{r + z_0 - \lambda} \quad \lambda = \frac{r(\bar{\lambda} - 1)}{\bar{\lambda} + 1} = z_0, \tag{8.12}$$

where r and z_0 are shown in Figure 8.12.

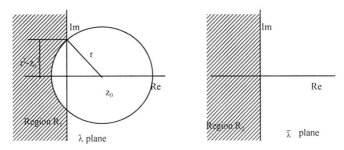

Fig. 8.12 The λ and $\bar{\lambda}$ planes.

Equation (8.11) can be mapped using Eq. (8.12) so that if the eigenval-
ues of the $\bar{\lambda}$ system are in the left half of the $\bar{\lambda}$ plane then the eigenvalues
of the λ system are in region R_1. If **A** is a non-negative definite matrix we
write $\mathbf{A} \geq \mathbf{0}$. A sufficient condition that the **F** matrix results in shifting the
eigenvalues of the closed loop system in the region R_1 is that the two ma-
trices in Eq. (8.13) be non-negative definite.

$$\overline{\mathbf{D}}(\mathbf{F}, r, z_0) = -(r^2 - z_0^2)\mathbf{I} + z_0\mathbf{D}(\mathbf{F}) + \Lambda^2 \geq 0,$$

$$\overline{\mathbf{K}}(\mathbf{F}, r, z_0) = -(r - z_0)^2\mathbf{I} - (r - z_0)\mathbf{D}(\mathbf{F}) + \Lambda^2 \geq 0.$$

(8.13)

The two matrices are obtained by mapping 8.11 into the $\overline{\lambda}$ plane. Such an F is called feasible. The system described by Eq. (8.11) is stable by its very structure. The problem is that the low-frequency natural frequency may have very small real parts. If we could find a matrix F so that the eigenvalues λ of Eq. (8.12) were in the region R_1 in Figure 8.12 we would have guaranteed damping of low-frequency modes. A direct test for eigenvalues λ in to be in R_1 in Figure 8.12 is quite difficult but a test for the eigenvalues, $\overline{\lambda}$ in Figure 8.12 to be in the left half plane is simple. Hence the mapping in Eq. (8.12) is used, recognizing that Eq. (8.13) is in the form

$$\mathbf{G} + z_0\mathbf{BFB}^{\mathrm{T}} \geq 0 \quad g_{ii} = u_i^2 - r^2 \quad g_{ij} = 0,$$

$$\mathbf{H} - (r - z_0)\mathbf{BFB}^{\mathrm{T}} \geq 0 \quad h_{ii} = v_i^2 \quad h_{ij} = 0,$$

(8.14)

where u_i and v_i are scalars depending on the geometry as shown in Figure 8.13, and G and H are diagonal matrices. Conditions are given in [7] but the numerical test is simply to apply the QR decomposition to the matrix B, i.e. if

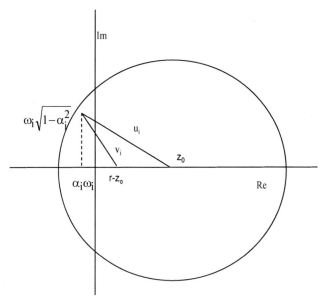

Fig. 8.13 The values of u_i and v_i for Eq. (8.14).

$$T^T B = \begin{bmatrix} R \\ 0 \end{bmatrix}, \tag{8.15}$$

where T is orthogonal and R is upper triangular. Then Eq. (8.14) is equivalent to

$$T^T G T + z_0 \begin{bmatrix} R \\ 0 \end{bmatrix} F \begin{bmatrix} R \\ 0 \end{bmatrix}^T = T^T G T + z_0 \begin{bmatrix} RFR^T & 0 \\ 0 & 0 \end{bmatrix} \geq 0,$$

$$\tag{8.16}$$

$$T^T H T - (r - z_0) \begin{bmatrix} R \\ 0 \end{bmatrix} F \begin{bmatrix} R \\ 0 \end{bmatrix}^T = T^T H T + (r - z_0) \begin{bmatrix} RFR^T & 0 \\ 0 & 0 \end{bmatrix} \geq 0.$$

The fact that G and H are diagonal, combined with the assumption that there are no open loop eigenvalues on the circle in Figure 8.11 allows one remaining transformation. If there are no eigenvalues on the circle, then M_{22} in Eq. (8.17) is invertible:

$$T^T G T = \begin{bmatrix} M_{11} & M_{12} \\ M_{12}^T & M_{22} \end{bmatrix}; \quad U = \begin{bmatrix} I & 0 \\ -M_{22}^{-1} M_{12}^T & I \end{bmatrix};$$

$$U^T T^T G T U = \begin{bmatrix} M_{11} - M_{12} M_{22}^{-1} M_{12}^T & 0 \\ 0 & M_{22} \end{bmatrix}. \tag{8.17}$$

If $T^T H T = N$, then the final conditions are that

$$M_{22} \geq 0$$

$$\frac{1}{r - z_0} R^{-1} \left[N_{11} - N_{12} N_{22}^{-1} N_{12}^T \right] R^{-T}$$

$$\geq F \tag{8.18}$$

$$\geq \frac{1}{z_0} R^{-1} \left[M_{11} - M_{12} M_{22}^{-1} M_{12}^T \right] R^{-T}$$

If there are sufficient well-placed measurements and there is one feasible matrix F then F is not unique. The choice of optimal F as the one with minimum Frobenius norm is suggested in [7]. The Frobenius norm of F is given by 8.19:

$$\|\mathbf{F}\| = \min_{\mathrm{F}}\left(\sum_{i=1}^{m}\sum_{j=1}^{m} f_{ij}^{2}\right). \tag{8.19}$$

In [7] it is shown that if the problem is one of minimizing the norm in Eq. (8.19) subject to the constraints of Eq. (8.18), one must compute the Schur factorization of

$$\frac{1}{z_0} R^{-1}\left[\mathbf{M}_{11} - \mathbf{M}_{12}\mathbf{M}_{22}^{-1}\mathbf{M}_{12}^{\mathrm{T}}\right]R^{-\mathrm{T}} = \mathbf{S}\mathbf{D}\mathbf{S}^{\mathrm{T}}. \tag{8.20}$$

In Eq. (8.20) $\mathbf{S}$ is orthonormal and $\mathbf{D}$ is diagonal. Replace the negative entries on the diagonal of $\mathbf{D}$ with zero and call the result $\overline{\mathbf{D}}$. If

$$\mathbf{S}\overline{\mathbf{D}}\mathbf{S}^{\mathrm{T}} \leq \frac{1}{r - z_0} R^{-1}\left[\mathbf{N}_{11} - \mathbf{N}_{12}\mathbf{N}_{22}^{-1}\mathbf{N}_{12}^{\mathrm{T}}\right]R^{-\mathrm{T}} \tag{8.21}$$

then $\mathbf{F} = \mathbf{S}\overline{\mathbf{D}}\mathbf{S}^{\mathrm{T}}$

The minimum norm $\mathbf{F}$ obtained from Eq. (8.21) may, in fact, not be feasible under some situations. Hence it is worth examining the closed loop eigenvalues before using it. The conditions in Eq. (8.18) provide an interval in which $\mathbf{F}$ must lie. The matrix in Eq. (8.18) is a more reliable choice for $\mathbf{F}$.

Example 8.5

The Matlab file TAC92.m has a 12-dimensional system with eigenvalues given in Table 8.3. The radius of the circle in Figure 8.3 is $r = 5$ and the center is at $z_0 = 4$. The B matrix is also shown in Table 8.3. Using $F = N_1$ the closed loop eigenvalues are shown in Figure 8.14.

Table 8.3 Data for Matlab example

Eigenvalues	B			Closed loop λs
$-0.1 \pm \mathrm{j}1$	3.6946	0.1730	0.1365	-9.3842
$-0.2 \pm \mathrm{j}2$	0.6213	1.9797	0.0118	-4.5731
$-0.3 \pm \mathrm{j}3$	0.7948	0.2714	2.8939	$-0.6803 \pm \mathrm{j}5.7594$
$-0.4 \pm \mathrm{j}4$	0.9568	0.2523	0.1991	$-0.6741 \pm \mathrm{j}4.7032$
$-0.5 \pm \mathrm{j}5$	0.5226	0.9757	0.2987	-0.4494 ± 3.9404
$-0.6 \pm \mathrm{j}6$	0.8801	0.7373	0.6614	Four at -1.000

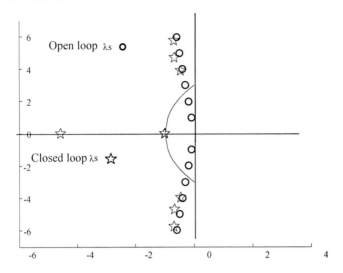

Fig. 8.14 Open and closed loop eigenvalues.

%TAC92.m 7-3-2007 Section 8.4

```
hold off
r=5;
z0=4;
w=[1 2 3 4 5 6];

B=[ 3.6946   0.1730   0.1365
    0.6213   1.9797   0.0118
    0.7948   0.2714   2.8939
    0.9568   0.2523   0.1991
    0.5226   0.8757   0.2987
    0.8801   0.7373   0.6614];

A(1:6,1:6)=0*eye(6);
A(1:6,7:12)=eye(6);
A(7:12,1:6)=-diag(w.^2);
A(7:12,7:12)=-2*diag(.1*w);

% plot open nlop eigenvalues and the circle
xp=-.1*[1:11]+.1;
plot(eig(A),'ro')
axis([-8 4 -8 8])
```

```
hold on
plot(xp,sqrt(9+8*xp-xp.*xp),xp,-sqrt(9+8*xp-xp.*xp))


%Form the diagonal matrices G and H is Eq. (8.14)
G=0*eye(6);
H=G;

for i=1:6,
   H(i,i)=w(i)^2-.2*w(i)+1;
   G(i,i)=0.8*w(1)+w(i)^2-9;
end

% now the QR decomposition of B and the formation of M and N in
% Eqs. (8.17) and (8.18)

[T,RR]=qr(B);
M=T'*G*T;
MM(1:3,1:3)=M(1:3,1:3)-M(1:3,4:6)*inv(M(4:6,4:6))*M(4:6,1:3);
MM(4:6,4:6)=M(4:6,4:6);
eig(MM(4:6,4:6));
N=T'*H*T;
NN(1:3,1:3)=N(1:3,1:3)-N(1:3,4:6)*inv(N(4:6,4:6))*N(4:6,1:3);
NN(4:6,4:6)=N(4:6,4:6);
%
R=RR(1:3,1:3);

% M1 and N1 are the matrices in equation (8.18)
M1=(1/z0)*inv(R)*MM(1:3,1:3)*inv(R)';
N1=(1/(r-z0))*inv(R)*NN(1:3,1:3)*inv(R)';

% the Schur decomposition for the minimum norm F
[S,V]=schur(M1);
VV=V;
for i=1:3,
if(V(i,i) < 0), VV(i,i)=0; end
end
F=S*VV*S';
%Test eig(F-M1)

F=N1;
eig(G+4*B*F*B')
```

eig(H-B*F*B')
%Plot Closed loop eigenvalues
A(7:12,7:12)=A(7:12,7:12)-B*F*B';
plot(eig(A),'bx')

Example 8.6

The system in Figure 8.15 [4] is a four-generator two-area system used as
an example of a power system stabilizer (PSS) design. The four generators
are identical except for the inertias which are 6.5 s in Area 1 and 6.175 in
area 2. The two areas are connected by two 220-km 230 kV lines. Without
the stabilizer there is an unstable low-frequency interarea mode at 0.64 Hz.
The system has been modified [8] by adding a parallel HVDC line connecting the two areas. In [8] an artificial situation is created to study the effect of collocated control using the HVDC line. The situation studied is a
period with the PSS not in service and after the 0.64-Hz oscillation has
grown the PSS is inserted to damp the oscillation. The power flow between
the systems is shown in Figure 8.16

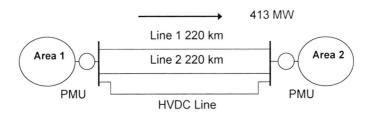

Fig. 8.15 The model two area system.

The HVDC line is modeled as two loads: one load connected to the bus at
each area interface. The amount of power drawn by one load is exactly
equal to the power supplied by the other load. The collocated control
scheme is to modulate the power flow along the line proportional to the
frequency difference between areas measured by the PMUs at either end of
the line. The frequency measurement taken at the remote bus (Area 2) is
subject to communications delay. Assuming a dedicated fiber-optic communication channel connecting Areas 1 and 2 it is reasonable to assume that
the delay is less than 50 ms. Figure 8.17 shows the collocated controller

performance with no delay. Delays up to 300 ms cause no noticeable degradation in the performance.

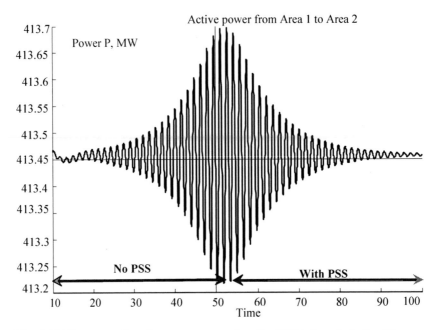

Fig. 8.16 The power flow between areas when the PSS is inserted after 50 s.

The compelling reason to use the collocated scheme is its robustness. To test the robustness the following system changes were made without changing the control algorithm.

1. Increase system-wide loading by 50%:
   A1: +50%,   A2: +50%
   A1: −50%,   A2: −50%
   A1: +50%,   A2: −50%
   A1: −50%,   A2: +50%

2. Increase nominal power flow along DC tie by 100 MW.
3. Change the lengths of the AC tie lines by 10%.
4 Change the generator inertias:
   Area 1: 7.5 s
   Area 2: 5.175 s

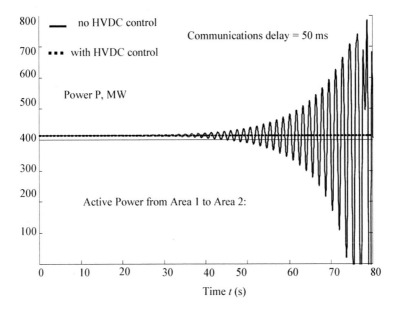

Fig. 8.17 Performance of the collocated controller.

For all these simulations the communications delay was kept at a constant 50 ms. None of these system changes had any noticeable effect on the performance of the controller.

8.5 Discrete event control

All of the control in the previous sections has been continuous feedback control, that is, the control, $\mathbf{u}(t)$, depends on the state, $\mathbf{x}(t)$, at each instant of time. It $\mathbf{x}(t)$ changes then $\mathbf{u}(t)$ changes with a small delay induced by communication latency (actually $\mathbf{u}(t)=\mathbf{f}(\mathbf{x}(t - \Delta t))$). The power system, however, has other types of control which can be characterized as discrete in their dependence on state. These controls are typically stability controls and are characterized by a specific action taken when the state exceeds a limit. Examples are the use of the dynamic brake, high-speed switching of series and shunt capacitors, high-speed generator tripping triggered by the loss of the direct current (DC) line, and under-frequency and voltage load shedding. The control action responds to the state but not continuously.

They are a form of discrete event control where state space has been partitioned by some process. In most early stability controls of this type many off-line simulations were performed in order to develop the rules for application of the control.

An early phasor measurement application was of this form [9]. The attempt was to control the power flow on the Intermountain and Pacific Intertie HVDC lines in a discrete form in response to a collection of approximately 20 phasor measurements which were to be communicated to the Sylmar substation (the southern terminal of both DC lines). The control action was limited to ± 500 MW ramps on each line. The measurement

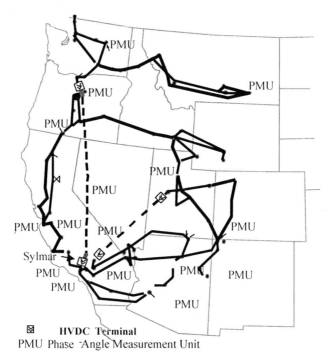

Fig. 8.18 The WECC System for DC line control.

locations are shown in Figure 8.18. The locations were chosen in part to attempt to get appropriate coverage of dynamic events but also in collaboration with the participating utilities in terms of accessibility of the PMUs and availability of communication channels. The power system model was a 176-bus system with 29 generators that included the two DC lines.

The challenge, from a control perspective, is to find a technique that will determine from the limited measurements which of the nine possibilities in

Table 8.4 is the correct choice. In attempting to select an approach it must be recognized that the off-line calculations of the classical approach are still appropriate; that is, the map from phasor measurements can be constructed from off-line simulations.

Table 8.4 Control options

Intermountain	Pacific Intertie
0	0
0	+500 MW
0	−500 MW
+500 MW	0
+500 MW	+500 MW
+500 MW	−500 MW
−500 MW	0
−500 MW	+500 MW
−500 MW	−500 MW

8.5.1 Decision trees

Consider the Xs and Os in the sequence of drawings in Figure 8.19. The horizontal and vertical lines partition the (x,y) plane into regions that are either Xs or Os. The horizontal line at $y = 1.5$ has all Xs below, the vertical line at $y = -0.5$ has all Xs to the left.

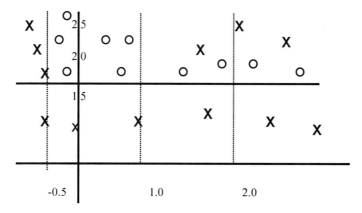

Fig. 8.19 Initial partitioning.

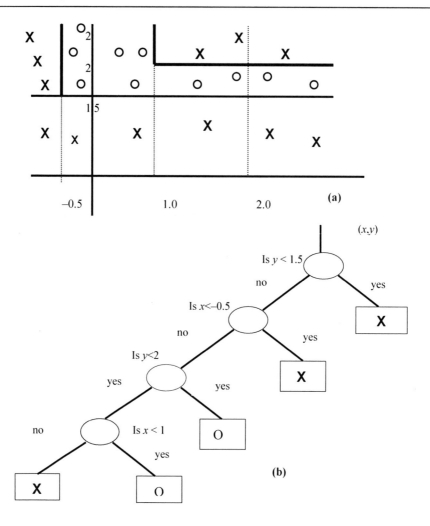

Fig. 8.20 (a) Recursive partitioning, and (b) the decision tree.

The partitioning is shown in Figures 8.19 and 8.20a,b with the decision tree produced by the process shown in Figure 8.20b. Such trees can be constructed by software packages for large databases [11]. A useful property of the decision tree training process is that if the process is given more data than is required to do the classification it will select the necessary inputs and discard the others. When finding the optimum PMU locations is an issue in almost all applications, the use of decision trees has the advantage of helping solve the placement problem automatically. A simple technique of allowing the training process to use more PMU data than will ultimately be practical can aid in determining the best locations.

The training data for the decision tree training involved thousands of four-second extended transient midterm stability program (ETMSP) transient stability runs [9]. Three-phase faults on all buses and transmission lines with fault durations from 0 s to 10 s were used to produce the training cases. The intended use of the tree logic is that the phasor measurements will be presented to the tree which will be able to decide which of the eight control actions to take. The first test is to see if the tree can successfully predict that an event will be unstable. The tree was 95% accurate in predicting stability/instability with the errors being on cases that were on the boundary between stability and instability. The tree training takes place off-line and is time consuming but the response of a trained tree is essentially limited by the delay in the arrival of the PMU data. This amounted to approximately 250 ms in the WECC application.

The fundamental requirement on the control is that it has a positive effect, that is, the control stabilized some unstable events but did not destabilize events that would be stable without control.

If post event generator angles for the first T seconds are denoted by $\delta_i(t)$ consider the objective function

$$F = \int_0^T \sum_i M_i (\delta_i(t) - \delta_{coa})^2 \, dt \tag{8.22}$$

where the Ms are machine inertias and δ_{coa} is the center of angle. The performance index F strongly penalizes diverging generator angles of large machines and provides the possibility of selecting control options that minimize F over a large number of initiating events. With a large number of machines and control options the computation is substantial but done off-line. Although the decision tree was trained as a stabilizing control (Eq. 8.22) and was not designed to control islanding, the resulting control would have prevented the December 14, 1995 event in which the WECC separated into five islands [10].

References

1. Stengel, R.F., "Stochastic Optimal Control: Theory and Application", John Wiley & Sons, New York, 1986.
2. Rostamkolai, N., Phadke, A.G., Thorp, J.S., and Long, W.F., "Measurement based optimal control of high voltage AC/DC systems", IEEE Transactions on Power Systems, Vol. 3, No. 3, August 1988, pp 1139–1145.
3. Manansala, E. C. and Phadke, A.G., "An optimal centralized controller with nonlinear voltage control", Electric Machines and Power Systems, 19, 1991, pp 139–156.

4. Kundur, P., "Power System Stability and Control", Example 12.6, p 813, McGraw-Hill, New York, 1994.
5. Smith, M. A., "Improved dynamic stability using FACTS devices with phasor measurement feedback", MS Thesis, Virginia Tech, 1994.
6. Mili, L. Baldwin, T., and Phadke, A.G., "Phasor measurements for voltage and transient stability monitoring and control", Workshop on Application of advanced mathematics to Power Systems, San Francisco, September 4–6, 1991.
7. Liu., J., Thorp, J.S., and Chiang, H-D, "Modal control of large flexible space structures using collocated actuators and sensors," IEEE Transactions on Automatic Control, Vol. 37, January, 1992, pp 143–147.
8. http://phasors.pnl.gov/Meetings/2007_may/presentations/synch_freq_meas.pdf Page 17 of the presentation
9. Rovnyak, S., Taylor, C. W., Mechenbier, J. R., and Thorp, J. S., "Plans to demonstrate decision tree control using phasor measurements for HVDC fast power changes," Conference on Fault and Disturbance Analysis and Precise Measurements in Power Systems, Arlington, VA, November 9, 1995.
10. Rovnyak, S., Taylor, C. W., and Thorp, J.S., "Real-time transient stability prediction – possibilities for on-line automatic database generation and classifier training," Second IFAC Symposium on Control of Power Plants and Power Systems, Cancun, Mexico, December 7, 1995.
11. http://www.salford-systems.com/cart.php

Chapter 9 Protection Systems with Phasor Inputs

9.1 Introduction

Synchronized phasor measurements have offered solutions to a number of vexing protection problems. These include the protection of series compensated lines, protection of multiterminal lines, and the inability to satisfactorily set out-of-step relays. In many situations the reliable measurement of a remote voltage or current on the same reference as local variables has made a substantial improvement in protection functions possible. In some examples communication of such measurements from one end of a protected line to the other is all that is required while in others communication across large distances is necessary.

Phasor measurements are particularly effective in improving protection functions which have relatively slow response times. For such protection functions, the latency of remote measurements is not a significant issue. For example, back-up protection functions of distance relays and protection functions concerned with managing angular or voltage stability of networks can benefit from remote measurements with propagation delays with latencies of up to several hundred milliseconds. Examples of applications of this nature are provided in Sections 9.4.

The next two sections will consider improved line protection using phasor measurements from the remote ends of the line. The following section involves adaptive protection in which the phasor measurements assist in "making adjustments automatically in various protection functions in order to make them more attuned to prevailing system conditions"[1,2].

9.2 Differential protection of transmission lines

Differential protection of buses, transformers, and generators is a well-established protection principle that has no direct counterpart in protection of long transmission lines. Pilot relays use communicated information

A.G. Phadke, J.S. Thorp, *Synchronized Phasor Measurements and Their Applications*,
DOI: 10.1007/978-0-387-76537-2_9, © Springer Science+Business Media, LLC 2008

from remote locations. True differential protection was not possible before synchronized phasor measurements. Communication over twisted pair of wires up to 5 miles is described in [2]. The advantages of differential protection are important for series compensated lines and tapped lines. There are a number of forms of current differentials for line protection. In the first form the currents are combined using a communication channel and compared. In the second form the currents are sampled and the samples communicated over a wide band channel, and in the third form phasors are computed from the samples and the phasor values communicated. The first is shown in Fig. 9.1.

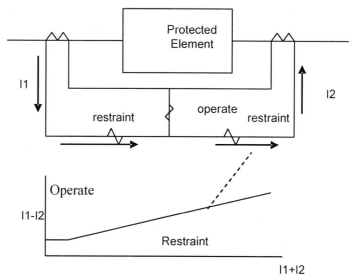

Fig. 9.1 Basic current differential.

The dashed dual slope shown in Fig. 9.1 is used for high-current conditions where current transformer (CT) accuracy and saturation is more likely. Transmission lines equipped with series compensation, flexible alternating current transmission system (FACTS) devices, or multiterminal lines present protection problems which call for differential protection. To date, such transmission line problems are solved with 'differential-like' schemes such as phase comparison. The easy availability of synchronized measurements using Global Positioning System (GPS) technology and the improvement in communication technology make it possible to consider true differential protection of transmission lines and cables.

Differential protection can be based on computed phasors or on samples, although it can be argued that significant shunt elements in the transmission line make phasors the preferred solution. In either case it is necessary to synchronize the sampling and time-tag the result. Phasors can be computed from fractional-cycle data windows as in impedance relaying, although full-cycle windows offer better security.

If I_i is the current phasor at terminal i (reference direction is positive when the current is flowing into the zone of protection), the differential currents may be defined as

$$|I_d| = |\sum_i I_i|$$ (9.1)

A single restraining current may be constructed by averaging the magnitudes of all terminal currents or taking the maximum of all the terminal currents as the restraint. Alternately, one restraining current for each pair of terminals may be constructed in order to maintain uniform sensitivity when one of the terminals of a multiterminal line is out of service. This is equivalent to the use of multiple restraints for multiwinding transformers.

If a two-terminal line is modeled with the exact-π equivalent [3] then the phasor currents and voltages are shown in Fig. 9.2.

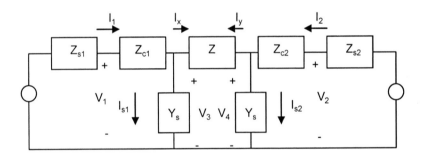

Fig. 9.2 Exact-π for the protected line.

The impedances Z_{c1} and Z_{c2} are the impedances of the possible series capacitor networks or FACTS devices, Z and Y_s are the exact-π impedance and admittance, respectively. If the relay measures I_1, V_1, I_2, and V_2, then the differential currents I_x and I_y can be obtained from Eqs. (9.2) and 9.3. Under no-fault conditions using Kirchhoff's current law $I_x = I_y$. When a fault occurs the 60-Hz exact-π is no longer valid because the currents and voltages are no longer pure fundamental frequency signals. A percentage differential characteristic such as shown in Fig. 9.1 based on I_x and I_y on a per-unit basis, with a modest slope, is capable of sensing faults within the zone defined by the terminal where I_x and I_y are measured.

$$V_3 = V_1 - I_1 Z_{c1}$$
$$V_4 = V_2 - I_2 Z_{c2}$$

(9.2)

$$I_{s1} = V_3 Y_s \quad I_{s2} = V_4 Y_s$$
$$I_x = I_1 - I_{s1} \quad I_y = I_2 - I_{s2}$$

(9.3)

The preceding discussion is for lines of any length because of the exact-π equivalent but has the disadvantage of requiring voltage measurements. In [4] an approximation to the charging current is proposed which does not require voltage measurement. The assumption is that each end uses data communicated from the other end to perform the current differential calculation.

The best synchronization is obviously obtained with GPS. Pre-fault load currents can also be used for synchronizing. Data communication over a dedicated fiber channel, while expensive, provides the best performance. A frequency shift power line carrier, voice-grade channel operating at 64 kbps, can also be used. The reliability of current differential schemes can be improved by adding redundant channels.

9.3 Distance relaying of multiterminal transmission lines

Occasionally lines are tapped without the benefit of a high side breaker as shown in Fig. 9.3. If there is no communication from terminal C then the first zone of the relay at A must always underreach terminals B and C; that is, zone 1 of the relay at A must be set with no infeed. If the three lines are all 10 ohms (secondary) and zone 1 is 85% of the line length then the relay at A has a 17-ohm zone 1 setting. In fact, only 70% of the line from the tap to B is in zone 1 for the relay at A. If the maximum infeed is such that $I_C = I_A$ then the impedance seen from A is 17 ohms when the fault is at 35% of the line from T to B. The relay reach is reduced but the alternative of setting zone 1 to 85% with infeed is that the relay could see faults beyond B without infeed. If there is no source at C then the load current I_C is in the opposite direction but is small compared to the fault current. The reach of the first zone at A would be extended by a very small amount by a modest outfeed (covered easily by the 85% setting).

The second zone at A must always overreach the complete lines AB and AC, however. This implies that zone 2 must be set with the infeed present.

Imagine the second zone at A is set at little greater than $1 + k$ of the line from A to B with the maximum infeed $I_C = I_A$ present. The multiplicative effect of the double current flowing from the point T to second zone setting point shown in Fig. 9.4 will give an impedance at A of $30 + 40\,k$ ohms. For example, if zone 2 is conventionally 150% ($k = 1/2$) the zone 2 impedance

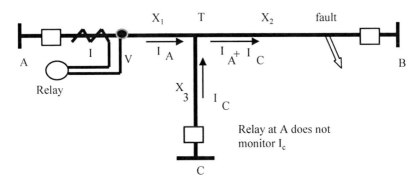

Fig. 9.3 A tapped transmission line.

at A is 50 ohms and could easily overreach zone 1. The compromise forced by the tapped line is that zone 2 must be held back. A value of k of 0.1 will give zone 2 = 34 ohms which could be difficult for a short line from B to D.

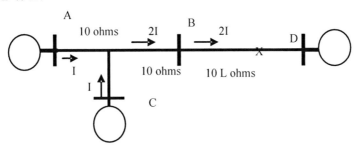

Fig. 9.4 Second zone setting for a tapped line.

Depending upon the status of breakers at B and C, the settings of table I or II are selected for relay at A as shown in Fig. 9.5.

The use of communication to signal the status of the breaker at C predates phasor measurements [5]. Adaptive schemes for setting zone 2 which do not use phasor measurements have also been reported [6]. The number of taps on a single line has increased over time, however, and the protection of a five-terminal line, for example, is far from simple. Schemes involving phasor measurement have been proposed which amount to differential relaying similar to bus protection.

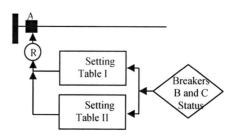

Fig. 9.5 Changing setting adaptively.

Software agents (a software agent is a computer program that takes independent action based on events in the surrounding environment [7]) have also been proposed to deal with these issues [8]. No practical agent-based systems have yet been reported in the literature.

9.4 Adaptive protection

Conventional protective systems respond to faults or abnormal events in a fixed, predetermined manner. This predetermined manner, embodied in the characteristics of the relays, is based upon certain assumptions made about the power system. "Adaptive relaying" accepts that relays may need to change their characteristics to suit prevailing power system conditions. With the advent of digital relays the concept of responding to system changes has taken on a new dimension. Digital relays have two important characteristics that make them vital to the adaptive relaying concept. Their functions are determined through software and they have a communication capability. This allows the software to be altered in response to higher-level supervisory software, under commands from a remote control center or in response to remote measurements.

Adaptive relaying with digital relays was introduced on a major scale in 1987 [1,2]. One of the driving forces that led to the introduction of adaptive relaying was the change in the power industry wherein the margins of operation were being reduced due to environmental and economic restraints and the emphasis was on operation for economic advantage. Consequently, the philosophy governing traditional protection and control performance and design have been challenged [9]. Adaptive protection is a protection philosophy which permits and seeks to make adjustments automatically in various protection functions in order to make them more attuned to prevailing system conditions. In 1993 a Working Group of the IEEE Power System Relaying Committee issued a report [10] with the results of a survey of relay engineers in North America questioning their acceptance of 16 specific adaptive functions and soliciting their suggestions for additional adaptive ideas. Examples include adapting transformer protection to the tap changer position, adaptive reclosing, and adapting relay characteristics to changes in load. It can be argued that adaptive relaying schemes address existing relaying deficiencies, making false trips less likely and improving the speed and dependability of the protection system. The result is an improvement in the reliability of the bulk power system and in some cases an increase in allowable power transfer limits. The adaptive relaying applications of interest in this chapter are those involving phasor measurements.

9.4.1 Adaptive out-of-step protection

It is recognized that a group of generators going out-of-step with the rest of the power system is often a precursor of a complete system collapse. Whether an electromechanical transient will lead to stable or unstable condition has to be determined reliably before appropriate control action could be taken to bring the power system to a viable steady state. Out-of-step relays are designed to perform this detection and also to take appropriate tripping and blocking decisions.

Traditional out-of-step relays use impedance relay zones to determine whether or not an electromechanical swing will lead to instability. A brief description of these relays and the procedure for determining their settings is provided here. In order to determine the settings of these relays it is necessary to run a large number of transient stability simulations for various loading conditions and credible contingencies. Using the apparent impedance trajectories observed at locations near the electrical center of the system during these simulation studies, two zones of an impedance relay are set, so that the inner zone is not penetrated by any stable swing. This is illustrated in the Fig. 9.6 (which uses reactance type of relay characteristics).

The outer zone is shown by a dashed line, and the inner zone is shown by a double line. Note that all the stable swing trajectories (shown by dotted lines) remain outside the inner zone, while all the unstable swing trajectories penetrate the outer as well as the inner zone. Although only two impedance characteristics are shown for stable and unstable cases, in reality a large number of such impedance loci must be examined. The time duration for which the unstable swings dwell between the outer and inner zones are identified as T_1 and T_2 for the two unstable characteristics shown in the figure. The largest of these dwell times (with an added margin) is chosen as the timer setting for the out-of-step relay. If an actual observed impedance locus penetrates the outer zone, but does not penetrate the inner zone before the timer expires, the swing is declared to be a stable swing. If it penetrates the outer zone and then the inner zone before the timer runs out, it is an unstable swing. Stable swings do not require any control action, whereas unstable swings usually lead to out-of-step blocking and tripping actions at predetermined locations.

Problems with traditional out-of-step relaysTraditional out-of-step relays are found to be unsatisfactory in highly interconnected power networks. This is because the conditions assumed when the relay characteristics are determined become out-of-date rather quickly, and in reality the electromechanical swings that do occur are quite different from those studied when the relays are set. The result is that traditional out-of-step relays often misoperate: they fail to determine correctly whether or not an evolving electromechanical swing is stable or unstable. Consequently, their control actions also are often erroneous, exacerbating the evolving cascading phenomena and perhaps leading to an even greater catastrophe. Wide-area measurements of positive-sequence voltages at networks (and hence swing angles) provide a direct path to determining stability using real-time data instead of using precalculated relay settings. This problem is very difficult to solve in a completely general case. However, progress could be made toward an out-of-step relay which adapts itself to changing system conditions. Angular swings could be observed directly, and time-series expansions could be used to predict the outcome of an evolving swing. It is highly desirable to develop this technique initially for known points of separation in the system. This is often known from past experience, and use should be made of this information. In time, as experience with this first version of the adaptive out-of-step relay is gained, more complex system structures with unknown paths of separation could be tackled.

It should be noted that a related approach was developed for a field trial at the Florida–Georgia interface [11–14] where the interface was modeled as a two-machine system. The machines in Fig. 9.6 are equivalents of the eastern interconnection on the left and Florida on the right with the four buses being physical buses in the interconnection.

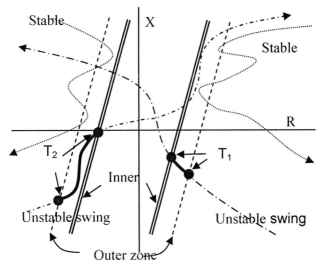

Fig. 9.6 Traditional out-of-step relay parameters using reactance-type relays and timers.

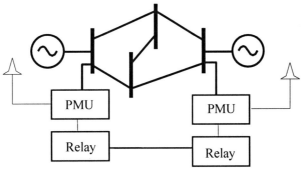

Fig. 9.7 Reduced Florida–Georgia system.

The equation of motion of the angle difference between the two rotors of the two machines is given by Eq. (9.4), where $\delta = \delta_1 - \delta_2$. M_1 and M_2 are the two rotor inertias, and the remaining terms in Eq. (9.4) are obtained from the equivalent system [15]. As the system undergoes changes due to a fault and its clearing, the parameters of the differential equation P_c and P_{max} change, and the classical equal-area criterion can be used to determine stability; that is, the area $A1$ must be smaller than the area $A2$ for stability. The issue in adaptive out-of-step is to determine the new parameter values P_c and P_{max} from real-time measurements. A least-squares estimate of P_m [15] from samples of δ is used in [11]. The estimate is obtained from five or six consecutive measurements of δ.

$$M \frac{d^2 \delta}{dt^2} = P_m + \{P_c - P_{max} \sin(\delta - \gamma)\} \qquad (9.4)$$

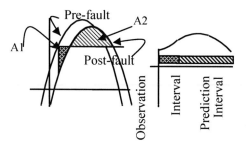

Fig. 9.8 The equal-area criterion.

Determination of coherent groups of machine

An algorithm can be developed for determining the principal coherent groups of machines as the electromechanical swings begin to evolve. Algorithms for inferring rotor angles from observed bus angles are needed. Criteria for judging coherency between machines and groups of machines will be developed. It is expected that centers of angles for each coherent group will be used in determining out-of-step condition.

Predicting the out-of-step condition from real time data

It is of course possible to determine whether or not a swing is unstable by waiting long enough and observing the actual swing. However, in order to take appropriate control action it is essential that a reliable prediction algorithm be developed which provides the stable–unstable classification of an evolving swing in a reasonable time. In the Florida–Georgia experiment a period of observation of actual angular swings for a maximum of 250 ms was used to obtain a reliable prediction of the outcome. Assuming that the normal periods of power system swings on a large interconnected power system are of the order of a few seconds, this target is reasonable. Experiments were conducted on the test system to determine what is the minimum observation period needed to predict the swing outcome with a chosen degree of confidence. With the observed swing evolution, a time-series approximation to the swings will be made in order to provide the predicted regions of the swings [11–14].

9.4.2 Security versus dependability

The existing protection systems are designed to be dependable at the cost of somewhat reduced security. This is a desirable bias when the power system is in a 'normal' state, meaning that there is sufficient operational margin in generation and transmission capability. The consequence of not tripping in primary protection time when a fault occurs in such cases is catastrophic in that transient instability and system collapse are likely to result. However, when the power system is in a stressed state, this is an unacceptable bias. Under stressed system conditions a false trip (insecure operation of the protection system) is likely to cause greater damage to the system. It is then desirable to alter the bias of the protection system in favor of increased security with a slightly increased possibility that the primary protection would not work as designed in case of a fault.

It should be recognized that a relay has two failure modes. It can trip when it should not trip (a false trip) or it can fail to trip when it should trip. The two types of reliability have been designated as "security" and "dependability" by protection engineers. Dependability is defined as the measure of the certainty that the relays will operate correctly for all faults for which they are designed to operate, while security is the measure of the certainty that the relays will not operate incorrectly. The existing protection systems with their multiple zones of protection and redundant systems are biased toward dependability, that is, a fault is always cleared by some relays. There are typically multiple primary protection systems often relying on different principles (one might depend on communications while another uses only local information) and multiple back-up systems that trip (with some time delay) if all primary systems fail to trip. The result is a system that virtually always clears the fault but as a consequence permits larger numbers of false trips. High dependability is recognized as being a desirable protection principle when the power system is in a normal "healthy" state, and high-speed fault clearing is highly desirable in order to avoid instabilities in the network. The consequent price paid in occasional false trip is an acceptable risk under "system normal" conditions. However, when the system is highly stressed false trips exacerbate disturbances and lead to cascading events.

An attractive solution is to "adapt" the security–dependability balance in response to changing system conditions as determined by real-time phasor measurements. The concept of "adaptive relaying" accepts that relays may need to change their characteristics to suit the prevailing power system conditions. The ability to change a relay characteristic or setting, on the fly, as it were, raised serious questions about reliability and responsibility. Adaptive relaying with digital relays was introduced on a major scale in 1987 [1,2]. One of the driving forces that led to the introduction of adaptive relaying was the change in the power industry wherein the margins of operation were being reduced due to environmental and

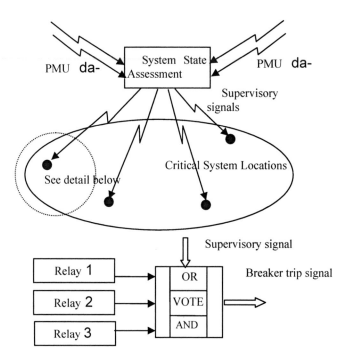

Fig. 9.9 Adjustment of dependability–security balance under stressed system conditions.

economic constraints and the emphasis was on operation for economic advantage. With three primary digital protection systems it is possible to implement an adaptive security–dependability scheme by using voting logic (see Fig. 9.9). The conventional arrangement is that if any of the three relays sees a fault then the breaker is tripped. A more secure decision would be made by requiring that two of the three relays see a fault before the trip signal is sent to the breaker. The benefit is in avoiding cascading and creating a more reliable system. The price paid for this increased security under "stressed" system conditions is that there is a somewhat reduced dependability, which is acceptable. The advantage of the adaptive voting scheme is that the actual relays are not modified but only the tripping logic responds to system conditions.

With three primary digital protection systems it is possible to implement an adaptive security–dependability scheme by using voting logic (see Fig. 9.9.). The conventional arrangement is that if any of the three relays sees a fault then the breaker is tripped. A more secure decision would be made by requiring that two of the three relays see a fault before the trip signal is sent to the breaker. The benefit is in avoiding cascading and creating a more reliable system. The price paid for this increased security under 'stressed' system conditions is that there is a somewhat reduced dependability, which is acceptable. The advantage of the adaptive voting scheme is that the actual relays are not modified but only the tripping logic responds to system conditions.

9.4.3 Transformer

Adaptive transformer protection gained immediate acceptance since it required only local information as opposed to adaptive schemes for other equipment. The slope of the characteristic in Fig. 9.1 is chosen to account for a variety of problems including CT ratio mismatches, saturation in CTs and transformers, and off-nominal turns ratios in tap-changing transformers. Slopes as great as 40% exist in some transformer relays. The price for a large slope is that part winding faults may fall on the wrong side of the tripping characteristic. Recognizing that, in a digital transformer relay, the sum and difference in the currents in Figure 9.1 are formed from samples, an adaptive solution to the off-nominal turns ratios is to monitor the tap changer and modify the trip and restraint currents appropriately. Similarly a record of the currents in no-fault conditions can be used to determine the actual CT ratios.

9.4.4 Adaptive system restoration

It must be accepted that some blackouts are unavoidable. It then becomes essential that strategies for restoring power after a blackout with minimum delays and at minimum cost should be put in place. Quick restoration of power is of paramount importance as it can significantly minimize user inconvenience due to power outages. Although precalculated restoration strategies obtained from planning-type simulation studies are available at present, they are often inadequate because the actual system state is quite different from the one assumed in the planning-type studies. Real-time wide-area measurements provide an excellent opportunity to determine a restoration strategy which takes into account the prevailing state of the power system [16].Although automated restoration procedures are possible to implement, for various practical reasons it is desirable to use a cooperative restoration technique, whereby the computer program suggests a restoration plan for any islands that may have been created and blacked-out, and for reconnecting the islands after they have been energized. The operator implements the suggested restoration plan based upon the step-by-step procedure provided by the computer program.

An example of how real-time data provided by phasor measurement units (PMUs) may have helped in restoration in the European blackout in 2003 is given in [17]. Review of the sequence of events showed that phase angle information was not known when operators were attempting to restore the initial line outage. Another example is a recent disturbance in Europe on November 4, 2006 [18]. Figure 9.10 shows PMU readings for reclosing attempts between two areas, including the final successful reclosing between those two areas and eventually with the third area. Restoration could only be achieved after the phase angles of separated areas became acceptable. In the absence of real-time data, seven unsuccessful attempts for restoration were made, ultimately leading to success when the angles were favorable. As Fig. 9.10 shows, had real-time data been available to the system operators, the unsuccessful attempts for restoration could have been avoided.

A recent study used "artificial neural networks" (ANNs) to accomplish system restoration [16]. In this proposed technique, the ANNs are trained for determining island boundaries for restoration in each viable island, and then determining a sequence of switching operations which would lead to restoration taking into account cold-load pick-up and control of overvoltages due to light load conditions.

If this approach is followed, then wide-area measurements could be used to determine prevailing angle differences across breakers used to close tie-lines between islands, and closing only when the angle differences are within acceptable bounds. If the angles exceed acceptable limits, a generation-load rescheduling within the islands would be implemented which would bring the angles within limits. This type of restoration scheme formalizes the lessons learned from uncontrolled restoration attempts as in the example of Fig. 9.10 which may lead to multiple failures on restoration attempts.

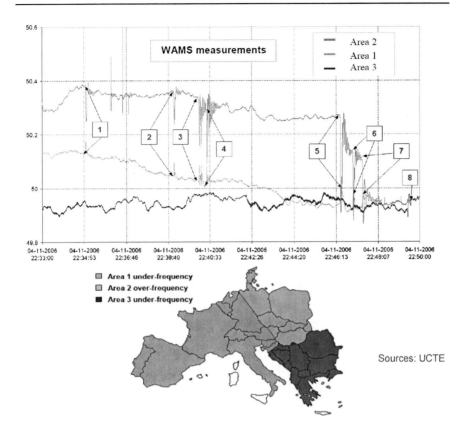

Fig. 9.10 PMU measurements from three areas during reclosing attempts, UCTE disturbance November 4, 2006.

9.5 Control of backup relay performance

It is well known that some back-up zones of distance relays are prone to tripping due to load encroachment during power system disturbances (see Fig. 9.11). This has led to a call for abandoning the use of back-up zones – in particular zone 3 of distance relays which is used to protect downstream circuits in case their protection systems fail to remove a fault on those circuits [19]. However, it has also been argued that this measure is too drastic, and should not be applied as a blanket policy. The remote back-up policy is designed to cover certain contingencies [20] for which no other protection is available. Under these circumstances, it becomes necessary to consider ways in which the loadability limits imposed by the remote back-up zones can be circumvented [21].

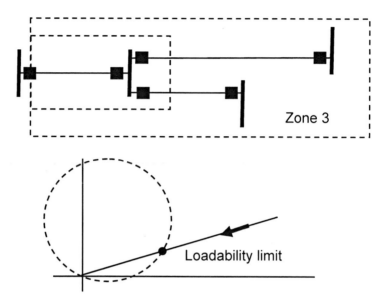

Fig. 9.11 Loadability limit imposed by a zone 3 setting of a distance relay. The illustration shows a mho characteristic, which is commonly used in many relays. As the load increased along the bold arrow, it would enter the tripping zone of the relay and cause an inappropriate trip.

Wide-area measurements offer a possibility for restraining the remote back-up relays in the event that the loading is being interpreted by the relay as a fault [22]. Consider the conditions illustrated in Fig. 9.11. Zone 3 of relay A is assumed to be picked up. If a significant negative-sequence current is present (indicating an unbalanced fault), the zone 3 pick-up is appropriate, and no further action is necessary. However, if the currents in the line are balanced, either a three-phase fault on the neighboring circuits or a possible loadability violation may be inferred. The PMUs at the buses corresponding to the terminals of lines which are to be backed-up by relay A may then determine if any of them see a zone 1 three-phase fault. This can be readily determined by taking the ratio of the positive-sequence voltage and current in those line terminals. If none of the PMUs indicate that a zone 1 three-phase fault exists, then the zone 3 pick-up of relay A must be due to loadability limit violation. If tripping on this condition by relay A is to be avoided, it would then be possible to block its operation by supervisory control of its output.

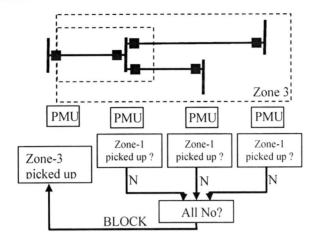

Fig. 9.12 Hidden failure monitoring and control.

9.5.1 Hidden failures

In examining the data made available by the North American Electric Reliability Council (NERC) [22] the extent of relay involvement in major disturbances becomes obvious. The mechanism has been referred to as "hidden failures" in the protection system [23]. It is not that relays initiate major disturbances but they tend to be involved in the spreading of what might have been a more localized event. The NERC data is an annual record of approximately 10 major disturbances measured by size and duration. Over a considerable period relays have played a role in about two-thirds of these events.

There are thousands of relays that operate correctly in any major event but if there is some defect in a relay the stressed system conditions that exist in a major disturbance can cause an incorrect relay operation. The fact that the defect is not noticed until conditions around it are unusual prompts the term "hidden". Maintenance can be a source of these hidden failures. This is true in power systems, the internet, and in large chemical plants. The entire process has been described as the "curse of robustness". Large complex system is designed to keep working when most of the elements are healthy. But under exceptional stress all the defective elements give way and the disturbance cascades to an exceptional extent [23].

Some defects in relays would cause the relay to misoperate immediately and do not qualify as "hidden". The exact mechanism of hidden failures in a number of commonly used protection schemes has been tabulated

[24–28]. Using this analysis it is possible to define the "regions of vulnerability" for hidden failures. The region of vulnerability for a given relay is the region in the power system where a fault will expose a hidden failure. To illustrate the concept we will assume the reach settings for relays that operate for faults within a certain distance of their location as shown in Table 9.1. An example of the region of vulnerability is shown in Fig. 9.13 for a variety of pilot schemes listed in Table 9.2. The region is the local reverse bus region where faults behind bus A will result in the circuit breaker at B tripping incorrectly for a fault in the shaded region. The mechanism of the failure is shown in Table 9.2.

Table 9.1 Relay setting distances

Type	Distance
Underreaching	0.7 of the line length
Overreaching	1.5 of the line length
Zone 2	1.2 of the line length
Zone 3	The entire line plus 1.2 of the length of the longest line The remote bus behind

Table 9.2 Hidden failures in pilot relays in Fig. 9.1.

Relay	Hidden failure mode
Directional comparison blocking	Fault detector at A cannot pick Up – transmitter fails to transmit
Directional comparison unblocking	DA continuously picks up
Permissive overreaching transfer trip	DA continuously picks up
Permissive underreaching transfer trip	Transmitter continuously transmits

The hidden failure probability is small and certainly is a function of loading and system conditions. The mechanism of hidden failure can be studied by considering a sample path made up of a sequence of hidden failure trips and correct trips due to overloads with resulting load and generation shedding. The probabilities are small enough that multiple hidden failure trips at one branch almost never happen. The sample path is approximately one-dimensional like a crack as opposed to a forest fire [29]. The hidden failure mechanism itself is sufficient to produce power-law behavior. The simulation presented in [29] produced the characteristic plot of the log of

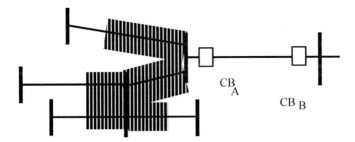

Fig. 9.13 Reverse local bus vulnerability region.

the size of the disturbance versus the log of the relative frequency of the disturbance as shown in Fig. 9.14. If two quantities are related by $y = x^{\alpha}$ the log–log plot similar to Fig. 9.14 is a straight line with slope α: that is, $\log y = \alpha \log x$

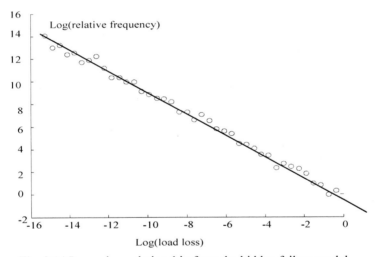

Fig. 9.14 Power-law relationship from the hidden failure model.

There are a number of papers and approaches to investigating this power-law mechanism [30]. The approach in [29] has been dubbed the hidden failure model.

One approach to reducing hidden failures or reducing their impact is to upgrade the relays at strategic locations determined with engineering judgement or simulation techniques. The upgrade is done with an eye toward hidden failures with an emphasis on self-monitoring and checking. The voting technique in Section 9.4.2 is also a possible tool.

9.6 Intelligent islanding

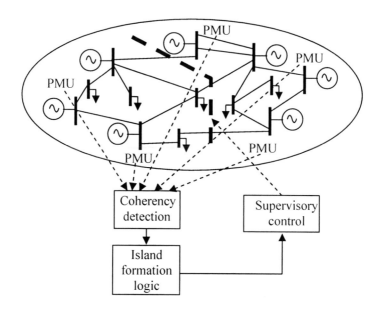

Fig. 9.15 Islanding based on coherency detection.

Separation of a power system into islands is a measure of last resort when the power system is stressed (thermal limits, phase angles across system centers beyond acceptable limits, voltage or frequency excursions beyond planned thresholds) and further disturbance propagation resulting in system separation is unavoidable. In principle, each resulting island should have a balance between generation and load in the island. In practice this may not be the case and consequently load or generation shedding may be required to bring about a balance and return the island to stable operation at normal frequency and voltage profile.

System separation into islands is accomplished using "system integrity protection scheme" (SIPS) also known as remedial action schemes (RASs) or system protection schemes (SPSs) [31]. These schemes are designed based on extensive planning studies covering various reasonable loading levels, topology, planned and unplanned outages, etc. In many practical situations the prevailing system conditions are quite different from those upon which the SIPS settings are based. Consequently, the performance of these systems may not be optimal for the prevailing system state.

Wide-area real-time synchronized data provide important information on prevailing system conditions to improve the match between SIPS and the actual system state. These measurements may be used to either supplement or replace precalculated scenarios and improve planned system separation in two key areas:

1. Using real-time data provided by the PMUs to more accurately determine whether a power system is heading to an unstable state and if a network separation is necessary to avoid a blackout.
2. Determine optimal islanding boundaries according to the prevailing system conditions. For example, establish which groups of generators will separate due to loss of synchronism and how to optimally balance load and generation in each island formed by coherent generator groups and loads.

Assuming that the PMU measurements are to be added to the existing SIPS plans to improve and speed up instability detection, technical, and computational requirements are well within the scope of present technology. The implementation would require a certain number of PMU measurements from optimally placed locations, and dedicated fiber optic channels with data latency of the order of 50 ms. The needed coherency detection algorithms and self-sufficient island identification algorithms would have to be developed to suit a specific power system.

In the absence of prior experience with the prevailing power system state, PMU-based SIPS would lead to islanding operations which are more appropriate for the existing system state. Wide-area measurement-based decisions made based upon real-time data should result in islanding operations which would form sustainable islands and better prospects for service restoration.

9.7 Supervisory load shedding

Underfrequency load shedding and restoration are used in most power systems to manage frequency excursions when islands of mismatched generation and load are formed. The frequency measurements are performed locally in distribution substations, and preassigned feeders tripped when frequency passes through preset trigger points [32]. More recently, with the threat of voltage collapse in many power systems, voltage-controlled load shedding has also been implemented [33,34]. Voltage collapse is a localized phenomenon, although several other factors (such as reactive power margins in generators) should also be taken into account. This is achieved by using SIPSs, with information brought from remote sites.

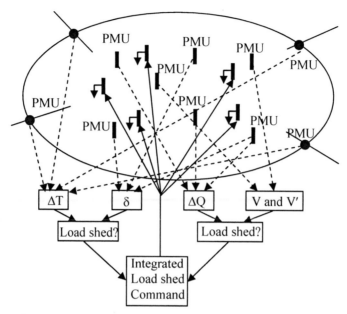

Fig. 9.16 Conceptual integrated load shedding based upon wide-area measurement systems.

SIPS systems are a form of wide-area measurement-based protection systems. It is possible to formulate a strategy which would address the is-sue of load shedding before the frequency begins to decay, or before the voltage begins a dive toward instability. A conceptual view of such a scheme is shown in Fig. 9.16. One could consider measuring a real-time "area control error" (ACE) by determining the deviation of tie-line power flows from their pre-disturbance schedule. While the tie-lines remain con-nected there is no frequency decay so that the ACE measures directly a shortfall of generation in the network. This shortfall can be weighed against predetermined allowable margins to retain stability and line load-ings below their capability. (The margin calculations could be made adap-tive to prevailing system conditions by using trained ANNs which would produce a decision as to whether or not load shedding is to be invoked. The ANN training could be performed using tie-flow deviations and bus phase angles at critical locations.) Should the thresholds be breached, one could initiate load shedding under centrally directed supervisory control. The required wide-area measurements for such a load-shedding scheme would be tie-line flows and phase angles at key network buses.

Similar strategy could also be employed for load shedding for voltage control. Wide-area measurements of voltages at key network buses would be collected at a central location, along with available amounts of reactive support in reactive power sources. Based upon this information, and using rate of change of voltage magnitudes at key buses, it would be possible to determine margin to voltage collapse and through it the need for load shedding. The advantage of such a scheme would be to consider the voltage problem in its entirety for the power system, and determine appropriate amounts of load to be shed in a coordinated fashion. In fact, the load shedding for real-power limitations (angular instability or overloads) and reactive power limitations (voltage profile violations or insufficient margins to voltage collapse) could be unified under an integrated load shedding control as shown in Fig. 9.16.

References

1. Horowitz, S.H., Phadke, A.G., and Thorp, J. S., "Adaptive transmission system relaying", IEEE Transactions Power Delivery, Vol. 3, No. 4, October 1988, pp 1436–1445.
2. Rockefeller, G.D., Wagner, C.L., Linders, J.R., Hicks, K.L., and Rizi, D.T., "Adaptive transmission relaying concepts for improved performance", IEEE Transactions, Vol. 3, No. 4, October 1988, pp 1446–1458.
3. Stevenson, W.D., "Elements of power system analysis," McGraw-Hill, New York, 1980
4. ABB Application manual, Line differential protection IED RED 670 ANSI
5. Hedding, R.A., and Mekic, F., "Advanced multi-terminal line current differential relaying and applications", Protective Relay Engineers, 60th Annual Conference , March 2007, pp 102–109.
6. Sidhu, T.S., Baltazar, D.S., Palomino, R.M., and Sachdev, M.S. "A new approach for calculating zone-2 setting of distance relays and its use in an adaptive protection system", IEEE Trans Power Delivery, Vol. 10, No.1, January 2004, pp 70–77..
7. Genesereth, M. and Ketchpel, S, "Software agents", Communications of the ACM 37(7), 48–52, 147, 1994.
8. Coury, D., Thorp, J., Hopkinson, K., and Birman, K., "Improving the protection of EHV teed feeders using local agents", Developments in Power System Protection, Conference Publication, No. 479, IEE 2001.
9. Thorp, J.S., Phadke, A.G., Horowitz, S.H., and Begovic, M.M., "Some applications of phasor measurements to adaptive protection, IEEE Transactions on PAS, Vol. 3, No. 2, May 1988, pp 791–798.
10. Thorp, J.S., et al, "Feasibility of adaptive protection and control", IEEE Transactions Power Delivery, Vol. 8, No. 3, July 1993, pp 975–983.
11. Centeno, V., Phadke, A.G., Edris, A., Benton, J., Gaudi, M., and Michel, G., "An adaptive out-of-step relay [for power system protection]", IEEE Transactions Power Delivery, Vol. 12, No. 1, January 1997 pp 61–71.

12. Centeno, V., Phadke, A.G., and Edris, A., "Adaptive out-of-step relay with phasor measurement", Developments in Power System Protection, Sixth International Conference on (Conf. Pub l. No. 434), 25–27 March 1997, pp 210–213.
13. Centeno, V., Phadke, A.G., Edris, A., Benton, J., and Michel, G., An adaptive out-of-step relay, IEEE Power Engineering Review, Vol. 17, No. 1, January 1997, pp 39–40.
14. Centeno, V., de la Ree, J., Phadke, A.G., Michel, G., Murphy, R.J., and Burnett, R.O., Jr., "Adaptive out-of-step relaying using phasor measurement techniques, Computer Applications in Power, IEEE Vol. 6, No. 4, October 1993, pp 12–17.
15. Phadke, A.G. and Thorp, J.S., Computer Relaying for Power Systems, Research Studies Press, Somerset, England 1988.
16. Bretas, A.S. and Phadke, A.G., "Artificial neural networks in power system restoration", IEEE Trans. On Power Delivery, Vol. 18, No. 4, October 2003, pp 1181–1186.
17. System Disturbance on November 4, 2006, UCTE, February 2007, available at www.ucte.com
18. Developments in UCTE and Switzerland, W. Sattinger, WAMC course at ETH Zürich, 30 August–1 September 2005.
19. Horowitz, S.H., and Phadke, A.G., "Boosting immunity to blackouts", Power and Energy Magazine, IEEE Vol. 1, No. 5, September–October 2003, pp 47–53.
20. Horowitz, S.H., and Phadke, A.G., "Third zone revisited", Power Delivery, IEEE Transactions on Vol. 21, No. 1, January 2006, pp 23–29.
21. Phadke, A.G., Novosel, D., and Horowitz, S.H., "Wide area measurement applications in functionally integrated power systems", CIGRE B-5 Colloquium, Madrid, Spain, 2007.
22. http://www.nerc.com/~dawg/dawg-disturbancereports.html
23. Taylor, C.W., "Improving grid behavior", IEEE Spectrum ,Vol. 36, No. 6, June 1999, pp 40–45.
24. Tamronglak, S., Horowitz, S.H., Phadke, A.G., and Thorp, J.S., Anatomy of power system blackouts: preventive relaying strategies. Power Delivery, IEEE Transactions on, Vol. 11, No. 2, April 1996, pp 708–715.
25. Elizondo, D.C., De La Ree, J., Analysis of hidden failures of protection schemes in large interconnected power systems, Power Engineering Society General Meeting, 2004. IEEE, Vol. 1, 6–10 June 2004, pp 107–114.
26. De La Ree, J., and Elizondo, D.C., "A methodology to assess the impact of hidden failures in protection schemes", Power Systems Conference and Exposition, 2004. IEEE PES, vol. 3, 10–13 October 2004, pp 1782–1783.
27. Phadke, A.G., and Thorp, J.S.,"Expose hidden failures to prevent cascading outages in power systems", Computer Applications in Power, IEEE, Vol. 9, No. 3, July 1996, pp 20–23.
28. Elizondo, D.C., De La Ree, J., Phadke, A.G., and Horowitz, S., Hidden failures in protection systems and their impact on wide-area disturbances, Power

Engineering Society Winter Meeting, 2001. IEEE, Vol. 2, 28 January-1 Febraury 2001, pp 710–714.

29. Wang, H. and Thorp, J.S., "Optimal locations for protection system enhancement: a simulation of cascading outages," IEEE Trans. on Power Delivery, Vol. 16, No. 4, October 2001, pp 528–33.

30. Dobson, I., Chen, J., Thorp, J.S., Carreras, B.A., and Newman, D.E., "Examining criticality of blackouts in power system models with cascading events," Proceedings of the 35th Annual Hawaii International Conference on System Sciences, January 2002.

31. Horowitz, S.H. and Phadke, A.G., "Power System Relaying", (book), Third Edition, John Wiley & Sons, RSP, 2008.

32. Westinghouse, "Applied Protective Relaying", (book), Westinghouse Electric Corporation, Newark, N.J., 1976, Chapter 19.

33. Begovic, M., Novosel, D., Karlsson, D., Henville, C., and Michel, G., "Wide-Area Protection and Emergency Control", Proceedings of the IEEE, Vol. 93, No. 5, May 2005, pp 876–891.

34. Madani, V., Novosel, D., Apostolov, A., and Corsi, S., " Innovative solutions for preventing wide area disturbance propagation," IREP Symposium for Bulk Power Systems Dynamics and Control VI, Cortina d'Ampezzo, Italy, August 2004.

Chapter 10 Electromechanical Wave Propagation

10.1 Introduction

Several different events connected with the early application of phasor measurements prompted consideration of the propagation of transient events in power systems. The first is typical of what is shown in Figure 10.1

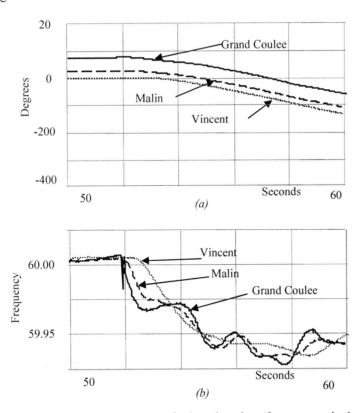

Fig. 10.1 Phasor measurements in three locations for an event in the WECC.

A.G. Phadke, J.S. Thorp, *Synchronized Phasor Measurements and Their Applications*,
DOI: 10.1007/978-0-387-76537-2_10, © Springer Science+Business Media, LLC 2008

It can be seen that there seems to be a delay between the frequencies at different points in the system. Similar effects were noticed in a variety of experiments [1, 2]. The event described in [1] was a load rejection test in Texas monitored with PMUs. The authors state 'Note the delay detecting the transient between the point closest to the plant (Venus) and the furthest (Robinson). There is nearly a half second delay between the onset of the frequency disturbance near the plant and its appearance at a similar level at a remote site. The propagation phenomenon is not clear. It is not electrical in nature because of the time lag. It appears to be related to the localized electrical inertia in the system.' A similar staged event in July, 1995 showed a delay of approximately a second between PMUs in Florida and New York [3].

The second motivation was a desire to display the phasor measurements obtained in the study described in Section 8.5.1 for the WECC 1994 disturbance. The plot of Figure 10.2 was produced by locating the phasor measurements geographically on a map of the WECC, making the z variable at that point the angle measurement, and then fitting a smooth surface to those points. By sampling the phase angles in time a movie of the surface can be obtained. Constant contour lines are shown below on a map showing state borders and the location of the two DC lines. Since to a first approximation power flows down hill in angle, transmission lines should be constructed perpendicular to the constant contour lines. Of course, the contour map changes in response to system conditions

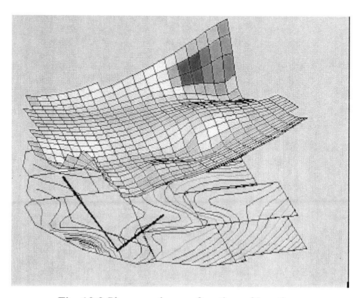

Fig. 10.2 Phase angle as a function of location.

The movies produced show wave-like motion of the surface when simulated or measured phasor quantities are used to draw the surfaces. There have been early attempts to describe this behavior even before phasor measurement data were available. Over a 30-year period [4–7] a number of authors have derived a wave equation for an idealized continuum power system model. In [4–7] linearized swing equations are written with the DC power-flow assumptions for a uniform, isotropic, and lossless network. A different approach in [7] produced an elliptic partial differential equation for a continuum load flow. In [7] the load flow equations are written under the DC load flow assumptions but the uniformity and isotropy constraints are relaxed. In [8–10] a combination of the two approaches produces a different model.

10.2 The Model

Consider the element at point (x, y) shown in Figure 10.3 with

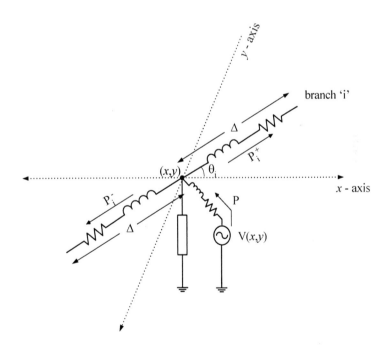

Fig. 10.3 Incremental power system model.

$R_i + jX_i$:   p.u. line impedance
$1/(R_s+jX_s)$:   p.u. generator admittance
G_s+jB_s: p.u. shunt admittance
$|V|$:   voltage magnitude
$\phi(x,y)$:   internal voltage phase angles
$\delta(x,y)$:   external voltage phase angles
M: generator inertia
D: generator damping
P: mechanical power injection
θ: angle of the branch with respect to the x–y axis

The power flow at the external node (x, y) must satisfy the load flow equation in Eq. (10.1)

$$
\begin{aligned}
\sum_{i=1}^{N} &\frac{\Delta R_i \left|V^2\right|}{\Delta^2(R_i^2 + X_i^2)}\left[1 - \cos(\delta(x, y) - \delta(x + \Delta x_i, y + \Delta y_i))\right] \\
&+ \frac{\Delta X_i \left|V^2\right|}{\Delta^2(R_i^2 + X_i^2)}\left[\sin(\delta(x, y) - \delta(x + \Delta x_i, y + \Delta y_i))\right] \\
&= \frac{\Delta R_s \left|V^2\right|}{\Delta^2(R_s^2 + X_s^2)}\left[1 - \cos(\delta(x, y) - \varphi(x, y))\right] \\
&+ \frac{\Delta X_s \left|V^2\right|}{\Delta^2(R_s^2 + X_s^2)}\left[\sin(\delta(x, y) - \varphi(x, y))\right]
\end{aligned}
\tag{10.1}
$$

And the power flow at the internal node(x,y) must satisfy the swing equation in Eq. (10.2). The next step is to write Taylor series expansions of $\delta(x \pm \Delta x_i, y \pm \Delta y_i)$, to use trigonometric identities for small angles and take the limit as $\Delta \to 0$. Two coupled partial differential equations result. One is the swing equation in Eq. (10.3) where virtually everything is a function of x and y, that is, $m(x,y)$, $d(x,y)$, $p_m(x,y)$, $g_{int}(x,y)$, $b_{int}(x,y)$, $\phi(x,y)$, and $\delta(x,y)$. The other equation is the continuum load flow equation in Eq. (10.4). The conductance **G** and susceptance **B** are 2×2 tensor fields, functions of (x,y) which capture the nonuniformity and anisotropy in the network

$$M\frac{\partial^2 \phi}{\partial t^2} + D\frac{\partial \phi}{\partial t} = P -$$

$$\frac{\Delta X_s |V^2|}{\Delta^2 (R_s^2 + X_s^2)}[1 - \cos(\phi(x,y) - \delta(x,y))] \tag{10.2}$$

$$\frac{\Delta X_s |V^2|}{\Delta^2 (R_s^2 + X_s^2)}[\sin(\varphi(x,y) - \delta(x,y))]$$

$$m\frac{\partial^2 \phi}{\partial t^2} + d\frac{\partial \phi}{\partial t} = p_m - g_{int}[1 - \cos(\phi - \delta)]] - b_{int}\sin(\phi - \delta) \tag{10.3}$$

$$-\nabla \bullet [\mathbf{B}(\nabla \delta)] + \nabla \delta \bullet \mathbf{G} \bullet \nabla \delta =$$

$$g_{int}[\cos(\delta - \phi) - 1] - b_{int}[\sin(\delta - \phi)] - g_s \tag{10.4}$$

Equations (10.3) and (10.4) are coupled by the angle difference $\delta - \phi$. Equation (10.4) is a continuum load flow with no time dependence. In principle, given a $\phi(x,y)$, equation (10.4) can be solved for $\delta(x,y)$. Alternately, given $\phi(x,y)$ equation (10.3) can be solved for $\delta(x,y,t)$; that is, equation (10.3) is a swing equation for each (x,y).

The phase angle gradient field $\nabla \delta$ is imaged to the power-flow field P through

$$P = -\mathbf{B}(\nabla \delta) \tag{10.5}$$

Several things can be noticed. The first is that the power injected at a point (x,y) in the system is

$$p(x,y) = -\bar{\nabla} \bullet [\mathbf{B}(\bar{\nabla} \delta)] + \nabla \delta \bullet \mathbf{G} \bullet \nabla \delta \tag{10.6}$$

The nonlinearity is in the second term and is due to the electrical losses. The lossless models in [4, 5, 6, 7] had linear wave behavior because they had no losses. It must be observed that the equation in Eqs. (10.3) and (10.4) describe an electromechanical system, not an electromagnetic system. It may be disquieting that linear electrical losses produce a nonlinear effect in the electromechanical system. Further, the net power loss in a region R in the system is given by

$$P_R^{loss} = \iint_R (\overline{\nabla}\delta \bullet \mathbf{G} \bullet \overline{\nabla}\delta) dx dy \qquad (10.7)$$

A linear analogy can be made between the electromechanical model and the electromagnetic wave propagation model. It is summarized in Table 10.1

Table 10.1 Correspondance between linear electromagnetic and electromechanical systems.

Electromagnetism		Electromechanical	
Quantity	Relationship	Quantity	Relationship
Electric potential	$\overline{E} = -\overline{\nabla}V$	Voltage phase angle	$\overline{\Phi} = -\nabla\delta$
Electric field Intensity		Phase angle gradient field	
Permittivity tensor	$\overline{D} = \varepsilon\overline{E}$	Susceptance tensor	$\overline{P} = \mathbf{B}\,\Phi$
Electric flux density	$\rho = -\overline{\nabla}\overline{D}$	Power flow density	$p = -\overline{\nabla} \bullet \overline{P}$
Charge density		Power injection density	
$\rho = -\overline{\nabla} \bullet (\varepsilon\overline{\nabla}V)$		$p = -\overline{\nabla} \bullet (\mathbf{B}\overline{\nabla}\delta)$	

Example 10.1 A ring system.

As an example consider a ring system made up of 64 identical generators connected in a ring with identical transmission lines connecting the generator. If the equilibrium is chosen as a 2π increase in angle in the counterclockwise direction around the ring then power flows in the same direction (Figure 10.4). The discrete model has 64 differential and algebraic equations in the form of Eq. (10.8) or the continuum form is Eq. (10.9):

$$M_i \frac{d^2\phi_i}{dt^2} = P_i^m - b_{int}\left[\sin(\phi_i - \delta_i)\right]$$

$$\left[2 - \cos(\delta_i - \delta_{i+1}) - \cos(\delta_i - \delta_{i-1})\right] \qquad (10.8)$$

$$+b\left[\sin(\delta_i - \delta_{i+1}) + \sin(\delta_i - \delta_{i-1})\right] = b_{int}\left[\sin(\phi_i - \delta_i)\right]$$

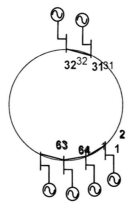

Fig. 10.4 Uniform power flow in a counterclockwise direction.

$$m(x)\frac{\partial^2 \varphi}{\partial t^2} = \left(\frac{2\pi}{64}\right)^2 - b_{int}\left[\sin(\phi(x) - \delta(x))\right] \tag{10.9}$$

$$\left(\frac{\partial \delta}{\partial x}\right)^2 - b\frac{\partial^2 \delta}{\partial x^2} = b_{int}\left[\sin(\phi(x) - \delta(x))\right] \tag{10.10}$$

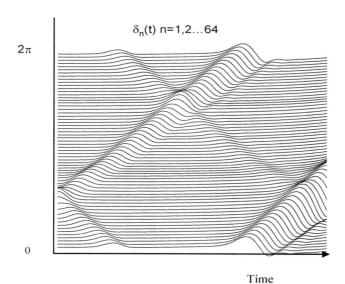

Fig. 10.5 The 64 angles from Example 10.1 versus time.

The model has no source admittance and hence no dispersion. The 64 angles from the discrete model are shown with δ_1 at the bottom and δ_{64} at the top. Figure 10.5 shows an initial perturbation from the equilibrium about a quarter of the way around the ring propagating in both directions. Remarkably it grows in the counterclockwise direction but decays in the clockwise direction.

Example 10.2

A more realistic power flow is around both sides of the ring in the same direction. Let the equilibrium angles be

$$\delta_{ek} = k\pi / 32, \, k = 1, 2, ... 32$$

$$\delta_{ek} = (64 - k)\pi / 32, \, k = 33, 34, ... 64 \qquad (10.11)$$

$$\text{with} \quad \delta = \frac{1}{2} e^{-0.1(x - 15.5)^2}$$

Then the waves are as shown in Figure 10.6

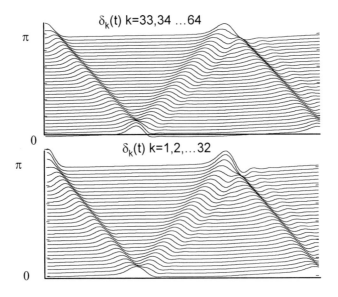

Fig. 10.6. Phase angles versus time for Example 10.2.

Vertical cross-sections of the phase angle for Example 10.2 are shown in Figure 10.7. Each of the eight figures is the angle at a fixed instant in time plotted versus position around the ring $\delta(\theta,t)$, where θ is the position on the ring as shown for the first cross-section. The pulse separates into two pulses in the second cross-section, spreads in the next two cross-sections, crosses and returns to the center in the last two cross-sections.

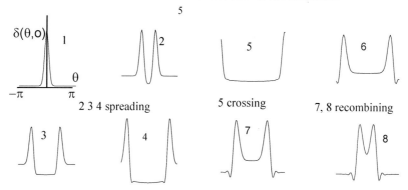

Fig. 10.7 Vertical cross sections of phase angle.

10.3 Electromechanical telegrapher's equation

If we open up the loop in Example 10.1 so we have a line rather than a ring, a system such as that shown in Figure 10.8 is produced.

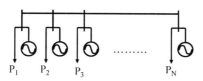

Fig. 10.8 A one-dimensional line.

If we assume small internal impedances and $R \ll X$ the equations become

$$m(x)\frac{\partial \omega}{\partial t} = -\frac{\partial p}{\partial x}$$

$$\frac{\partial p}{\partial t} = -b(x)\frac{\partial \omega}{\partial x},$$

$$(10.12)$$

where

$$P = -b\frac{\partial \delta}{\partial x} \quad \omega = \frac{\partial \delta}{\partial t}. \tag{10.13}$$

The result is a form of the Telegraphers equation in angle and power rather than voltage and currents.

$$m(x)\frac{\partial^2 \delta}{\partial t^2} = p_m(x) + b(x)\frac{\partial^2 \delta}{\partial x^2}, \tag{10.14}$$

which has

$$\text{velocity} = \sqrt{\frac{b(x)}{m(x)}}, \quad Z_0 = 1/\sqrt{b(x)m(x)}. \tag{10.15}$$

In other words the one-dimensional line has a characteristic velocity of propagation and a characteristic 'impedance'. The impedance is not in ohms, since the ratio is angle divided by power rather than volts divided by amps. The velocity of propagation measured in the power system is far less than the velocity of light and varies in different parts of the country. Speeds of hundreds of miles a second to a thousand miles per second have been observed in the frequency measurement network (FNET) system [11] where the measured waves are in frequency rather than angle. They are, of course, a different manifestation of the same waves.

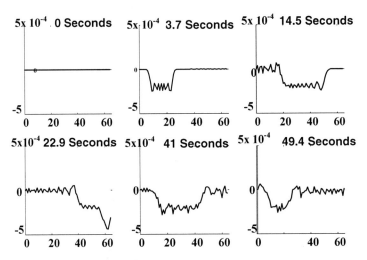

Fig. 10.9 Time snapshots of angular velocity versus position.

Determining the onset of a wave can be complicated with electromechanical waves just as in electromagnetic waves. The use of discriminant functions as in traveling wave relays [12] is possible

$$D_f = \omega + Z_0 p = 2\omega^+,$$
$$D_r = \omega - Z_0 p = 2\omega^-.$$

(10.16)

Typical plots of angular velocity versus position along a uniform line are shown in Figure 10.9 at a sequence of times. Although the waves begin crisply, they become distorted as time goes on. The forward and reverse discriminant functions are shown in Figure 10.10

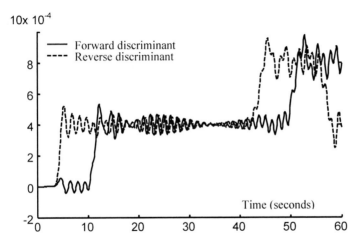

Fig. 10.10 Forward and reverse discriminate functions for the example.

10.4 Continuum voltage magnitude

The development of Section 10.2 assumed constant voltages. The load flow equation can be re-derived with the external bus voltage magnitude depending on (x,y), that is,

$$E(x, y)e^{j\delta(x,y)}.$$

(10.17)

The resulting two equations are a real power equation

$$p(x,y) = -\nabla \bullet E^2(B\overline{\nabla}\delta) - \overline{\nabla} \bullet E(G\overline{\nabla}E) + \overline{\nabla}E \bullet G \bullet \overline{\nabla}E + (E\overline{\nabla}\delta) \bullet G \bullet (E\overline{\nabla}\delta) \tag{10.18}$$

and a reactive power equation

$$q(x,y) = -\nabla \bullet E \; (B\overline{\nabla}E) - \overline{\nabla} \bullet E^2(G\overline{\nabla}\delta) + \overline{\nabla}E \bullet B \bullet \overline{\nabla}E + (E\overline{\nabla}\delta) \bullet B \bullet (E\overline{\nabla}\delta). \tag{10.19}$$

A striking simplification is possible if we assume the R/X ratio of the lines is constant , $\rho = R/X$, and use a transformation of real and reactive power introduced in [13]. Equation (10.20) is a rotation of the real and reactive power so that the new 'real power' is rotated by the line angle as shown in Figure 10.11:

$$\begin{bmatrix} \widetilde{p}(x,y) \\ \widetilde{q}(x,y) \end{bmatrix} = \frac{1}{1+\rho^2}\begin{bmatrix} 1 & -\rho \\ \rho & 1 \end{bmatrix}\begin{bmatrix} p(x,y) \\ q(x,y) \end{bmatrix} \tag{10.20}$$

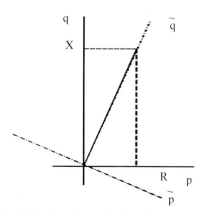

Fig. 10.11 Rotation of real and reactive power.

In the new coordinate system the two partial differential equations become

$$\widetilde{p}(x,y) = -\overline{\nabla} \bullet E^2(B\overline{\nabla}\delta)$$
$$\widetilde{q}(x,y) = -\overline{\nabla} \bullet E(B\overline{\nabla}E) + \overline{\nabla}E \bullet B \bullet \overline{\nabla}E + (E\overline{\nabla}\delta) \bullet B \bullet (E\overline{\nabla}\delta). \tag{10.21}$$

The angle dependence in the second equation can be removed and a single equation in the voltage magnitude written as in Eq. (10.22):

$$bE^3\left(\frac{\partial^2 E}{\partial x^2}+\frac{\partial^2 E}{\partial y^2}\right)+bE^2\tilde{q}(x,y)=\frac{b}{4}\left(\int_0^x \tilde{p}(\hat{x},y)\mathrm{d}\hat{x}+\int_0^x \tilde{p}(x,\hat{y})\mathrm{d}\hat{y}\right).(10.22)$$

Example 10.3 The one-dimensional line of Figure 10.7 with 40 sections has $p(x)$ and $q(x)$ shown in Figure 10.12. The continuous solution obtained from Eq. (10.22) along with the discrete solution are shown in Figure 10.13.

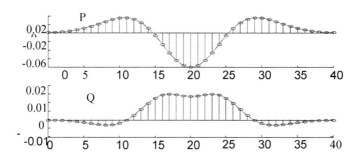

Fig. 10.12 Real and reactive power for the one-dimensional line.

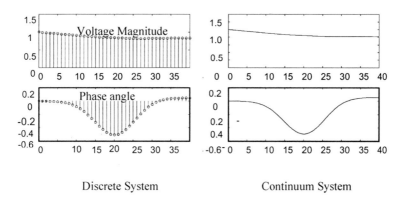

Discrete System   Continuum System

Fig. 10.13 Discrete and continuous solutions for the voltage on the one-dimensional line.

10.5 Effects on protection systems

One of the motivations for considering electromechanical waves in power system is the concern about these disturbances on protection systems. It is safe to say that relay systems were not designed with the motion of waves in angle and frequency propagating through the system at speeds of hundreds of miles a second. To investigate these effects the 64-machine system was investigated with each line protected with an overcurrent relay, a distance relay, an out-of-step relay, and a load shedding relay [14]. In theory and as observed on the system, electromechanical waves travel at speeds slow enough to make it possible to communicate the existence of the wave to remote locations before the wave arrives. The diagram in Figure 10.14 is a simple model of a monitoring system for such a scheme.

10.5.1 Overcurrent relays

The overcurrent relays are set to pickup at twice maximum load. The wave propagation is generated by applying a pulse with a 0.5 radian peak value to the 16th machine. In addition at the 16th machine two other lies show possible overcurrent violations. The current waveforms for lines at buses 5 and 36 are shown in Figure 10.15.

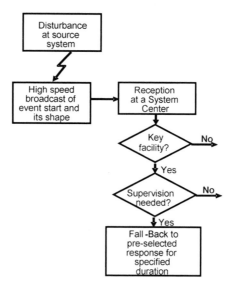

Fig. 10.14 Wave propagation monitoring and control system.

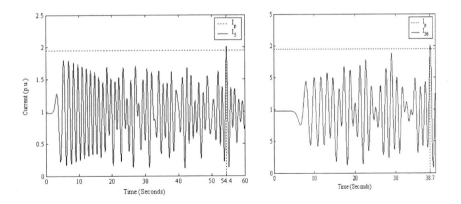

Fig. 10.15 Overcurrent relay pickup setting and transmission line current.

10.5.2 Impedance relays

The distance relays were set for zone 1 at 90% of the line length, zone 2 at 150% of the line length, and zone 3 at 150% of the next line length. A similar wave was initiated with a 1.5 radian peak at machine 16. Zone 1 of the relay on the adjacent line on bus 15 was entered by the apparent impedance from the wave as shown in Figure 10.16.

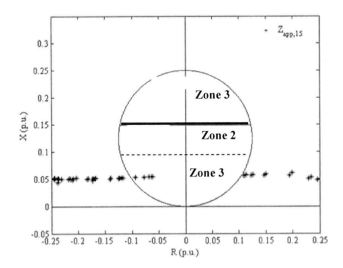

Fig. 10.16 Partial locus of apparent impedance movement at bus 15.

10.5.3 Out-of-step relays

The inner zone of the relay is set at $j0.2$ pu and the outer zone chosen as $j0.25$ pu. The timer setting to determine whether the trajectory is a fault or an unstable swing is 0.02 s. A second timer used to distinguish between a stable or unstable swing is sat at 0.16 s. A Gaussian disturbance with peak value of 2.5 radians is applied to the 16th machine at $t = 5$ s in order to generate the disturbances.

When the disturbance propagates through the ring system, out-of-step relays located at the 14th bus through the 18th bus are entered as shown in Figure 10.17, and some relays are in danger of tripping due to the wave propagation. Since wave propagation is a transient phenomenon, it is desirable to block the tripping of the out-of-step relays for the duration of the wave propagation.

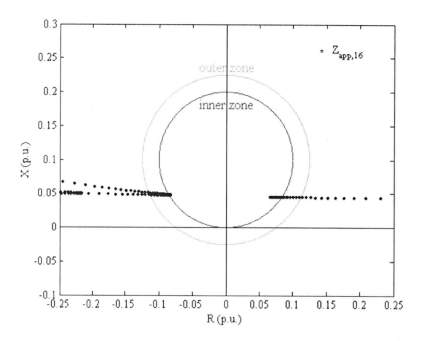

Fig. 10.17 Partial locus of apparent impedance movement at the 16th bus ($t_1 = 0.16$ s, $t_2 = 0.07$ s, $t_3 = 0.12$ s, $t_4 = 0.11$ s, and $t_5 = 0.12$ s).

10.5.4 Load shedding

For load shedding the pickup setting was set at 59.8 Hz with the time delay of ten or more cycles. A Gaussian disturbance with peak value of two radians is applied to the 16th machine in order to generate the disturbances. Frequencies detected at the 15th, 16th, and 17th buses are below 59.8 Hz. The frequency at bus 16 is shown in Figure 10.18.

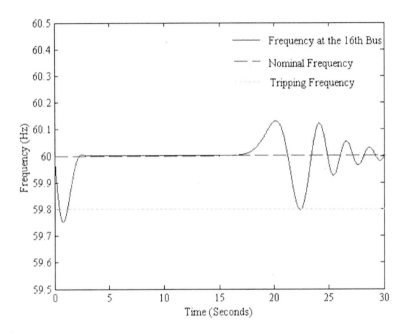

Fig. 10.18 Frequency at the 16th bus ($t_1 = 0.72$ s and $t_2 = 0.22$ s).

10.6 Dispersion

The term $\overline{\nabla}\delta \bullet G \bullet \overline{\nabla}\delta$ in Eq. (10.6) represents the dispersion in the waves. With the angle written as in Eq. (10.23):

$$\delta(x,t) = e^{kx - wt}, \tag{10.23}$$

where k is the wave number: there is negligible dispersion when

$$k \ll \sqrt{\beta/b} \quad \text{and significant dispersion when} \quad k \approx \sqrt{\beta/b}. \tag{10.24}$$

Most of the figures have had little dispersion for clarity but Figure 10.19 shows a wave with dispersion.

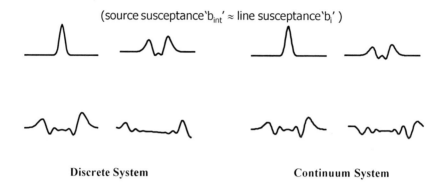

Fig. 10.19 Time snapshots of phase angle versus position.

Distortion in the waveforms is also found when the system is nonuniform. For example, when the inertias in the 64-machine ring system have a random component rather than being identical the cross-sections are distorted as in Figure 10.19.

10.7 Parameter distribution

There is certainly no expectation that control center software will be called upon to solve partial differential equations. The aim of this chapter has been to establish that electromechanical disturbances spread in the power system at speeds much less than the speed of light because of the effect of machine inertias. This observation opens the door to new concepts in adaptive protection and control and the possible prevention of blackouts. It is also a view of the system that might be appropriate for a regional coordinator rather than the operators in the ISO control center. In August of 2003 a view such as Figure 10.2 of Ohio might have been very useful in New York. Given this motivation, this section addresses the problem of smoothing the power system data so a picture like Figure 10.2 has some degree of accuracy.

Consider the one line diagram with a geographical base (where is bus 221 on a map?), the line data, the generator data, and load data; how do we create a continuum model? How do we convert a detailed model with thousands of point and lines into a smooth continuum such as the view of the power system from the space station? Consider a density function with the following properties

$$f(x) \geq 0, \quad \int_{\Omega} f(s)\mathrm{d}s = 1 \quad \partial^m f(x) \Big/ \partial x^m \qquad (10.25)$$

exists. (It is not required but a Gaussian density would be a good candidate.) Imagine convolving the data at a point (inertia at a bus, a load in megawatts) with the density. The inertia would become distributed in a Gaussian mound around the original point. The width of the distribution determines the accuracy and smoothness of the model. A very narrow distribution preserves everything but does not gain much. A very broad density will smooth out the system loosing precise details but gives a satellite view of the system. A symbolic version is shown in Figure 10.20.

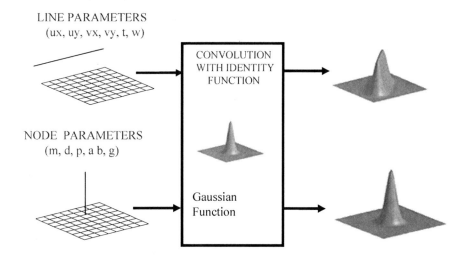

Fig. 10.20 Convolution with a density function.

Figure 10.21 shows a surface similar to Figure 10.2 for a typical load flow solution for the IEEE 118-bus system with the line structure below. The surface is relatively smooth as would be expected. It is possible a regional reliability coordinator or an operator might see relatively similar surfaces

on a daily or hourly basis so that unusual system conditions, when they occurred, would stand out – the possibility that surfaces like Figure 10.2
might become the 'face' of the system.

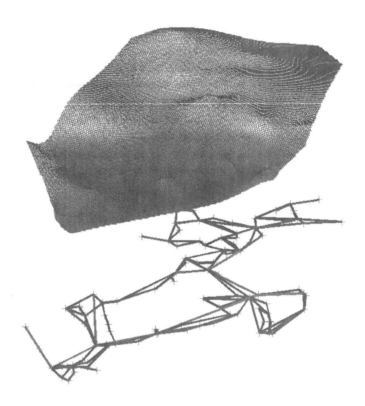

Fig. 10.21 The phase angle plot for the continuum 118-bus system with the line
structure[15].

References

1. Faulk, D. and Murphy, R.J., "Comanche peak unit No 2 100 percent load re-
    jection test – underfrequency and voltage phasors measured across TU electric
    system", Protective Relay Conference Texas A&M, March 1994.
2. Murphy, R.J., "Power disturbance monitoring", Western Protective Relaying
    Conference, 1995.
3. Murphy, R.J., Personal Communication.

4. Semlyen, A., "Analysis of disturbance propagation in power systems based on a homogeneous dynamic model", IEEE Transactions on PAS, Vol. 93, pp 676–684, March 1974.
5. Cresap, R.L. and Hauer, J.F., "Emergence of a new swing mode in the Western power system", IEEE Transactions on Power Apparatus and Systems, Vol. PAS-100, No. 4, April 1981, pp 2037–2045.
6. Grobovoy, A. and Lizalek, N., "Assessment of power system properties by wave approach and structure analysis", Fifth International Conference on Power System Management and Control, 2002, April 2002.
7. Dersin, P. and Lewis A.H., "Aggregate feasibility sets for large power networks" Proceedings of the 9th Triennial World Congress IFAC, Vol. 4, Budapest, Hungary, July 1984, pp 2163–2168.
8. Thorp, J.S., Seyler, C.E., and Phadke, A.G., "Electromechanical wave propagation in large electric power systems", IEEE Transactions on CAS, Vol. 45, No. 6, June 1998, pp 614–622.
9. Thorp, J.S., Seyler, C.E., Parashar, M., and Phadke, A.G., "The large scale electric power system as a distributed continuum", Power Engineering Letters, IEEE Power Engineering Review, January 1998, pp 49–50.
10. Parashar, M. and Thorp, J. S., "Continuum modeling of electromechanical dynamics in large-scale power systems", IEEE Transactions on, June 2004, pp 1851–1858.
11. Zhong, Z. et al., "Power system frequency monitoring network (FNET) implementation", IEEE Transactions on Power Systems, Vol. 20, No. 4, November 2005, pp. 1914–1920.
12. Dommel, H.W. and Michels, J.M., "High speed relaying using traveling wave transient analysis", IEEE paper No. A78-214-9.
13. Haque, M.H., "Novel decoupled load flow method", IEEE Proceedings-C, Vol. 140, No. 1, May 1993, pp 199–205.
14. Huang, L., Parashar, M., Phadke, A.G., and Thorp, J.S., "Impact of electromechanical wave propagation on power-system reliability", Proceedings of the 39th CIGRE Conference, Paris, France, August 2005.
15. Parashar, M. "Continuum modeling of electromechanical dynamics in power systems", Ph.D. Dissertation, Cornell University, 2003.

Index

Continued from page ii

Maintenance Scheduling in Restructured Power Systems
M. Shahidehpour and M. Marwali
ISBN 978-0-7923-7872-3

Power System Oscillations
Graham Rogers
ISBN 978-0-7923-7712-2

State Estimation in Electric Power Systems: A Generalized Approach
A. Monticelli
ISBN 978-0-7923-8519-6

Computational Auction Mechanisms for Restructured Power Industry Operations
Gerald B. Sheblé
ISBN 978-0-7923-8475-5

Analysis of Subsynchronous Resonance in Power Systems
K.R. Padiyar
ISBN 978-0-7923-8319-2

Power Systems Restructuring: Engineering and Economics
Marija Ilic, Francisco Galiana, and Lester Fink, eds.
ISBN 978-0-7923-8163-1

Cryogenic Operation of Silicon Power Devices
Ranbir Singh and B. Jayant Baliga
ISBN 978-0-7923-8157-0

Voltage Stability of Electric Power Systems
Thierry Van Cutsem and Costas Vournas
ISBN 978-0-7923-8139-6

Automatic Learning Techniques in Power Systems
Louis A. Wehenkel
ISBN 978-0-7923-8068-9

Energy Function Analysis for Power System Stability
M. A. Pai
ISBN 978-0-7923-9035-0

Electromagnetic Modelling of Power Electronic Converters
J. A. Ferreira
ISBN 978-0-7923-9034-3

Spot Pricing of Electricity
F. C. Schweppe, M. C. Caramanis, R. D. Tabors, R. E. Bohn
ISBN 978-0-89838-260-0

Breinigsville, PA USA
09 April 2010
235731BV00009B/70/P